T0210769

Statistical Tools for Environmental Quality Measurement

APPLIED ENVIRONMENTAL STATISTICS
CHAPMAN & HALL/CRC

Environmental Statistics with S-PLus
Steven P. Millard and Nagaraj K. Neerchal

Statistical Tools for Environmental Quality Measurement
Michael E. Ginevan and Douglas E. Splitstone

Statistical Tools for Environmental Quality Measurement

Michael E. Ginevan
Douglas E. Splitstone

CRC Press
Taylor & Francis Group
Boca Raton London New York

CRC Press is an imprint of the
Taylor & Francis Group, an **informa** business

CRC Press
Taylor & Francis Group
6000 Broken Sound Parkway NW, Suite 300
Boca Raton, FL 33487-2742

Visit the Taylor & Francis Web site at
http://www.taylorandfrancis.com

and the CRC Press Web site at
http://www.crcpress.com

Cover design by Jason Miller
Technical typesetting by Marilyn Flora

Library of Congress Card Number 2003055403

Library of Congress Cataloging-in-Publication Data

Ginevan, Michael E.
  Statistical tools for environmental quality measurement / Michael E. Ginevan.
    p. cm. — (Applied environmental statistics)
  Includes bibliographical references and index.
  ISBN 1-58488-157-7 (alk. paper)
    1. Environmental sciences—Statistical methods. I. Splitstone, Douglas E. II. Title. III.
Series.

GE45.S73G56 2003
363.7′064—dc22   2003055403

Table of Contents

Preface

Statistics is a subject of amazingly many uses and surprisingly
few effective practitioners. (Efron and Tibshirani, 1993)

The above provocative statement begins the book, *An Introduction to the Bootstrap*, by Efron and Tibshirani (1993). It perhaps states the truth about the traditional lament among the organized statistics profession. "Why aren't statisticians valued more in the practice of their profession?" This lament has been echoed for years in the addresses of presidents of the American Statistical Association, notably Donald Marquardt (*The Importance of Statisticians*, 1987), Robert Hogg (*How to Hope With Statistics*, 1989), and J. Stuart Hunter (*Statistics as a Profession*, 1994).

A clue as to why this lament continues can be found by spending a few hours reviewing the change over time from 1950 to the present in statistical journals such as the *Journal of the American Statistical Association* or *Technometrics*. The emphasis has gradually swung from using statistical design and reasoning to solve problems of practical importance to the consumers of statistics to that of solving important statistical problems. Along with this shift in emphasis the consumer of statistics, as well as many statisticians, have come to view the statistician as an oracle rather than a valuable assistant in making difficult decisions.

Boroto and Zahn (1989) captured the essence of the situation as follows:

> ... Consumers easily make distinctions between a journeyman statistician and a master statistician. The journeyman takes the problem the consumer presents, fits it into a convenient statistical conceptualization, and then presents it to the consumer. The journeyman prefers monologue to dialogue. The master statistician hears the problem from the consumer's viewpoint, discusses statistical solutions using the consumer's language and epistemology, and arrives at statistically based recommendations or conclusions using the conceptualizations of the consumer or new conceptualizations that have been collaboratively developed with the consumer. The master statistician relies on dialogue.[*]

---

[*] Reprinted with permission from *The American Statistician*. Copyright 1989 by the American Statistical Association. All rights reserved.

An Overview of This Book

The authors of the work are above all statistical consultants who make their living using statistics to assist in solving environmental problems. The reader of this text will be disappointed if the expectation is to find new solutions to statistical problems. What the reader will find is a discussion and suggestion of some statistical tools found useful in helping to solve environmental problems. In addition, the assumptions inherent in the journeyman application of various statistical techniques found in popular USEPA guidance documents and their potential impact on the decision-making process are discussed. The authors freely admit that the following chapters will include the occasional slight bending of statistical theory when necessary to facilitate the making of the difficult decision. We view this slight bending of statistical theory as preferable to ignoring possibly important data because they do not fit a preconceived statistical model.

In our view statistics is primarily concerned with asking quantitative questions about data. We might ask, "what is the central tendency of my data?" The answer to this question might involve calculation of the arithmetic mean, geometric mean, or median of the data, but each calculation answers a slightly different question. Similarly we might ask, "are the concentrations in one area different from those in another area?" Here we might do one of several different hypothesis tests, but again, each test will answer a slightly different question. In environmental decision-making, such subtleties can be of great importance. Thus in our discussions we belabor details and try to clearly identify the exact question a given procedure addresses. We cannot overstate the importance of clearly identifying *exactly* the question one wants to ask. Both of us have spent significant time redoing analyses that did not ask the right questions.

We also believe that, all else being equal, simple procedures with few assumptions are preferable to complex procedures with many assumptions. Thus we generally prefer nonparametric methods, which make few assumptions about the distribution of the data, to parametric tests that assume a specific distributional form for the data and may carry additional assumptions, such as variances being equal among groups. In some cases, such as calculation of upper bounds on arithmetic means, parametric procedures may behave very badly if their assumptions are not satisfied. In this regard we note that "robust" procedures, which will give pretty good answers even if their assumptions are not satisfied, are to be preferred to "optimal" procedures, which will work really well if their assumptions are satisfied, but which may work very badly if these assumptions are not satisfied.

Simplicity is to be preferred because at some point the person doing the statistics must explain what they have done to someone else. In this regard we urge all consumers of statistical analyses to demand a clear explanation of the questions posed in an analysis and the procedures used to answer these questions. There is no such thing as a meaningful analysis that is "too complex" to explain to a lay audience.

Finally, we cheerfully admit that the collection of techniques presented here is idiosyncratic in the sense that it is drawn from what, in our experience, "works."

Often our approach to a particular problem is one of several that might be applied (for example, testing "goodness of fit"). We also make no reference to any Bayesian procedures. This is not because we do not believe that they are useful. In some cases a Bayesian approach is clearly beneficial. However we do believe that Bayesian procedures are more complex to implement and explain than typical "frequentist" statistics, and that, in the absence of actual prior information, the benefits of a Bayesian approach are hard to identify. In some cases we simply ran out of time and room. Using multivariate statistics to identify the sources of environmental contamination is one area we think is important (and where a Bayesian approach is very useful) but one that is simply beyond the scope of this book. Watch for the second edition.

Chapter 1 discusses the often ignored but extremely important question of the relationship of the measurement taken to the decision that must be made. While much time and effort are routinely expended examining the adequacy of the field sampling and analytical procedures, very rarely is there any effort to examine whether the measurement result actually "supports" the decision-making process.

Chapter 2 provides a brief introduction of some basic summary statistics and statistical concepts and assumptions. This chapter is designed to assist the statistically naive reader understand basic statistical measure of central tendency and dispersion. The basics of testing statistical hypothesis for making comparisons against environmental standards and among sets of observations are considered in **Chapter 3**. **Chapter 4** discusses a widely used, but most misunderstood, statistical technique, *regression analysis*. Today's popular spreadsheet software supports linear regression analysis. Unfortunately, this permits its use by those who have little or no appreciation of its application, with sometimes disastrous consequences in decision making.

Tools for dealing with the nagging problem in environmental studies plagued by analytical results reported as below the limit of method detection or quantification are considered in **Chapter 5**. Most techniques for dealing with this "left censoring" rely upon an assumption regarding the underlying statistical distribution of the data. The introduction of the "empirical distribution function" in **Chapter 6** represents a relaxation in the reliance on assuming a mathematical form for the underlying statistical distribution of the data.

"Bootstrap" resampling, the subject of **Chapter 6**, at first glance seems to be a little dishonest. However, the basic assumption that the data arise as an independent sample representative of the statistical population about which inferences are desired is precisely the assumption underlying most statistical procedures. The advent of high-speed personal computers and the concept of bootstrap sampling provide a powerful tool for making inferences regarding environmentally important summary statistics.

Many environmentally important problems do not support the assumption of the statistical independence among observations that underlies the application of most popular statistical techniques. The problem of spatially correlated observations is discussed in **Chapter 7**. "Geostatistical" tools for identifying, describing, and using spatial correlation in estimating the extent of contamination and volume of contaminated material are discussed.

Chapter 8 considers the techniques for describing environmental observations that are related in time. These typically arise in the monitoring of ambient air quality, airborne and/or waterborne effluent concentrations.

Acknowledgments

We would be remiss if we did not acknowledge the contribution of our clients. They have been an unending source of challenging problems during the combined 60 plus years of our statistical consulting practice. CRC Press deserves recognition for their patience, as many deadlines were missed. We admire their fortitude in taking on this project by two authors whose interest in publishing is incidental to their primary livelihood.

A great vote of appreciation goes to those whose arms we twisted into reviewing various portions of this work. A particular thank-you goes to Evan Englund, Karen Fromme, and Bruce Mann for the comments and suggestions. All of their suggestions were helpful and thought provoking even though they might not have been implemented. Those who find the mathematics, particularly in Chapter 8, daunting can blame Bruce. However, we believe the reader will get something out of this material if they are willing to simply ignore the formulae.

A real hero in this effort is Lynn Flora, who has taken text and graphics from our often creative word-processing files to the final submission. It is due largely to Lynn's skill and knowledge of electronic publication that this book has been brought to press. Lynn's contribution to this effort cannot be overstated.

Finally, we need to acknowledge the patience of our wives, Jean and Diane, who probably thought we would never finish.

References

Boroto, D. R. and Zahn, D. A., 1989, "Promoting Statistics: On Becoming Valued and Utilized," *The American Statistician*, 43(2): 71–72.

Efron, B. and Tibshirani, R. J., 1998, *An Introduction to the Bootstrap*, Chapman & Hall/CRC, Boca Raton, FL, p. xiv.

Hogg, R. V., 1989, "How to Hope With Statistics," *Journal of the American Statistical Association*, 84(405): 1–5.

Hunter, J. S., 1994, "Statistics as a Profession," *Journal of the American Statistical Association*, 89(425): 1–6.

Marquardt, D. W., 1987, "The Importance of Statisticians," *Journal of the American Statistical Association*, 82(397): 1–7.

About the Authors

Michael E. Ginevan, Ph.D.

Dr. Ginevan, who received his Ph.D. in Mathematical Biology from the University of Kansas in 1976, has more than 25 years experience in the application of statistics and computer modeling to problems in public health and the environment. His interests include development of new statistical tools, models, and databases for estimating exposure in both human health and ecological risk analyses, development of improved bootstrap procedures for calculation of upper bounds on the mean of right skewed data, development of risk-based geostatistical approaches for planning the remediation of hazardous waste sites, computer modeling studies of indoor air exposure data, and analyses of occupational epidemiology data to evaluate health hazards in the workplace. He is the author of over 50 publications in the areas of statistics, computer modeling, epidemiology, and environmental studies.

Dr. Ginevan is presently a Vice President and Principal Scientist in Health and Environmental Statistics at Blasland, Bouck and Lee, Inc. Past positions include Leader of the Human Health Risk Analysis Group at Argonne National Laboratory, Principal Expert in Epidemiology and Biostatistics at the U.S. Nuclear Regulatory Commission, Deputy Director of the Office of Epidemiology and Health Surveillance at the U.S. Department of Energy, and Principal of M. E. Ginevan & Associates.

Dr. Ginevan is a founder and past Secretary of the American Statistical Association (ASA) Section on Statistics and the Environment, a recipient of the Section's Distinguished Achievement Medal, a past Program Chair of the ASA Conference on Radiation and Health, and a Charter Member of the Society for Risk Analysis. He has served on numerous review and program committees for ASA, the U.S. Department of Energy, the U.S. Nuclear Regulatory Commission, the National Institute of Occupational Safety and Health, the National Cancer Institute, and the U.S. Environmental Protection Agency, and was a member of the National Academy of Sciences Committee on Health Risks of the Ground Wave Emergency Network.

Douglas E. Splitstone

Douglas E. Splitstone, Principal of Splitstone & Associates, has more than 35 years of experience in the application of statistical tools to the solution of industrial and environmental problems. The clients of his statistical consulting practice include private industry, major law firms, and environmental consulting firms. He has designed sampling plans and conducted statistical analyses of data related to the extent of site contamination and remedial planning, industrial wastewater discharges, and the dispersion of airborne contaminants. He is experienced in the investigation of radiological as well as chemical analytes.

As a former manager in the Environmental Affairs Department for USX Corporation in Pittsburgh, PA, Mr. Splitstone managed a multi-disciplinary group of environmental specialists who were responsible for identifying the nature and cause of industrial emissions and developing cost-effective environmental control solutions. Mr. Splitstone also established statistical service groups devoted to environmental problem solution at Burlington Environmental, Inc., and the International Technology Corporation.

He has been a consultant to the USEPA's Science Advisory Board serving on the Air Toxics Monitoring Subcommittee; the Contaminated Sediments Science Plan review panel; and the Environmental Engineering Committee's Quality Management and Secondary Data Use Subcommittees. Mr. Splitstone is a member of the American Statistical Association (ASA) and is a founder and past chairman of that organization's Committee on Statistics and the Environment. He was awarded the Distinguished Achievement Medal by the ASA's Section on Statistics and the Environment in 1993.

Mr. Splitstone also holds membership in the Air and Waste Management Association, and the American Society for Quality. He has served as a technical reviewer for *Atmospheric Environment*, the *Journal of Official Statistics*, *Journal of the Air and Waste Management Association,* and *Environmental Science and Technology.* Mr. Splitstone received his M.S. in Mathematical Statistics from Iowa State University in 1967.

CHAPTER 1

Sample Support and Related Scale Issues in Sampling and Sampling Design[*]

> Failure to adequately define [sample] *support* has long been a source of confusion in site characterization and remediation because risk due to long-term exposure may involve areal supports of hundreds or thousands of square meters; removal by backhoe or front-end loader may involve minimum remediation units of 5 or 10 m^2; and sample measurements may be taken on soil cores only a few centimeters in diameter. (Englund and Heravi, 1994)

The importance of this observation cannot be overstated. It should be intuitive that a decision regarding the average contaminant concentration over one-half an acre could not be well made from a single kilogram sample of soil taken at a randomly chosen location within the plot. Obviously, a much more sound decision-making basis is to average the contaminant concentration results from a number of 1-kg samples taken from the plot. This of course assumes that the design of the sampling plan and the assay of the individual physical samples truly retain the "support" intended by the sampling design. It will be seen in the examples that follow that this may not be the case.

Olea (1991) offers this following formal definition of "support":

> An n-dimensional volume within which linear average values of a regionalized variable may be computed. The complete specification of the support includes the geometrical shape, size, and orientation of the volume. The support can be as small as a point or as large as the entire field. A change in any characteristic of the support defines a new regionalized variable. Changes in the regionalized variable resulting from alterations in the support can sometimes be related analytically.

While the reader contemplates this formal definition, the concept of sample support becomes more intuitive by attempting to discern precisely how the result of the sample assay relates to the quantity required for decision making. This includes reviewing all of the physical, chemical, and statistical assumptions linking the sample assay to the required decision quantity.

---

[*] This chapter is an expansion of Splitstone, D. E., "Sample Support and Related Scale Issues in Composite Sampling," *Environmental and Ecological Statistics,* 8, pp. 137–149, 2001, with permission of Kluwer Academic Publishers.

Actually, it makes sense to define two types of support. The desired "decision support" is the sample support required to reach the appropriate decision. Frequently, the desired decision support is that representing a reasonable "exposure unit" (for example, see USEPA, 1989, 1996a, and 1996b). The desired decision support could also be defined as a unit of soil volume conveniently handled by a backhoe, processed by incineration or containerized for future disposal. In any event, the "desired support" refers to that entity meaningful from a decision-making point of view. Hopefully, the sampling scheme employed is designed to estimate the concentration of samples having the "desired support."

The "actual support" refers to the support of the aliquot assayed and/or assay results averaged. Ideally, the decision support and the actual support are the same. However, in the author's experience, the ideal is rarely achieved. This is a very fundamental problem in environmental decision making.

Olea's definition indicates that it is sometimes possible to statistically link the actual support to the decision support when they are not the same. Tools to help with this linking are discussed in Chapters 7 and 8. However, in practice the information necessary to do so is rarely generated in environmental studies. While this may seem strange indeed to readers, it should be remembered that most environmental investigations are conducted without the benefit of well-thought-out statistical design.

Because this is a discussion of the issues associated with environmental decision making and sample support, it addresses the situation as it is, not what one would like it to be. Most statisticians reading this chapter would advocate the collection of multiple samples from a decision unit, thus permitting estimation of the variation of the average contaminant concentration within the decision unit and specification of the degree of confidence in the estimated average. Almost all of the environmental engineers and/or managers known to the authors think only in terms of the minimization of field collection, shipping, and analytical costs. Their immediate objective is to minimize the cost of site investigation and remediation. Therefore, the idea of "why take two when one will do" will usually win out over assessing the "goodness" of estimates of the average concentration.

This is particularly true in the private sector, which comprises this author's client base. If there is some potential to influence the design of the study (which is not a frequent occurrence), then it takes a great deal of persuasive power to convince the client to pay for any replicate sampling and/or assay. The statistician's choice, absent the power of design, is to either withdraw, or attempt to guide the decision-making process toward the correct interpretation of the results in light of the actual sample support.

If environmental investigators would adhere to the traditional elements of statistical design, the appropriate decisions would be made. These elements are nicely described by the U. S. Environmental Protection Agency's (USEPA) Data Quality Objectives Process (USEPA, 1994a; Neptune, 1990). Flatman and Yfantis (1996) provide a complete discussion of the issues.

The Story of the Stones

A graphic example of how the actual support of the assay result may be inconsistent with the desired decision support is provided by the story of the stones. In reality, it is an example of how an incomplete sampling design and application of standard sample processing and assay protocols can lead to biased results. This is the story of stone brought onto a site to facilitate the staging of site remediation. The site must remain confidential, however; identification of the site and actual data are not necessary to make the point.

Those who have witnessed the construction of a roadway or parking lot will be able to easily visualize the situation. To provide a base for a roadway and the remediation staging area, 2,000 tons of stone classified as No. 1 and No. 24 aggregate by the American Association of State Highway Transportation Officials (AASHTO) were brought onto the site. The nominal sizes for No. 1 and No. 24 stone aggregate are 3½ inches to 1½ inches and 2½ inches to ¾ inch, respectively. These are rather large stones. Their use at the site was to construct a roadway and remediation support area for trucks and equipment. In addition, 100 tons of AASHTO No. 57 aggregate stone were placed in the access roadway and support area as a top course of stone pavement. No. 57 aggregate has a nominal size of from 1 inch to No. 4 sieve. The opening of a No. 4 sieve is approximately 3/16 inch (see Figure 1.1).

Figure 1.1　Contrast between No. 57 and No. 1 Aggregate

Upon the completion of the cleanup effort for total DDT, the larger stone was to be removed from the site for use as fill elsewhere. Removal of the stone involves its raking into piles using rear-mounted rakes on a backhoe and loading via front-end loader into trucks for transport off-site. In order to remove the stone from the site, it had to be demonstrated that the average concentration of total DDT for the stone removed met the Land Disposal Restriction criterion of 87 microgram per kilogram ($\mu g/kg$).

The remedial contractor, realizing that the stone was brought on site "clean," and the only potential for contamination was incidental, suggested that two composite samples be taken. Each composite sample was formed in the field by combining stone from five separate randomly chosen locations in the roadway and support area. The total DDT concentrations reported for the two samples were 5.7 $\mu g/kg$ and 350 $\mu g/kg$, obviously not a completely satisfactory result from the perspective of one who wants to move the stone off-site.

It is instructive to look at what actually happened to the sample between collection and chemical assay. Because surface contamination was the only concern, the stones comprising each composite were not crushed. Instead several stones, described by the chemical laboratory as having an approximate diameter of 1.5 centimeters (cm), were selected from each composite until a total aliquot weight of about 30 grams was achieved. This is the prescribed weight of an aliquot of a sample submitted for the chemical assay of organic analytes. This resulted in a total of 14 stones in the sample having the 5.7-μg/kg result and 9 stones in the sample showing the 350-μg/kg result.

The stones actually assayed, being less than 0.6 inch (1.5 cm) in size, belong only to the No. 57 aggregate size fraction. They represent less than 5 percent of the stone placed at the site (100 tons versus 2,000 tons). In addition, it represents the fraction most likely to be left on site after raking. Thus, the support of the assayed subsample is totally different than that required for making the desired decision.

In this situation, any contamination of the stone by DDT must be a surface phenomenon. Assuming the density of limestone and a simple cylindrical geometric shape, the 350-μg/kg concentration translates into a surface concentration of 0.15 μg/cm^2. Cylindrical stones of approximately 4 cm in diameter and 4 cm in height with this same surface concentration would have a mass concentration of less than 87 μg/kg. Thus arguably, if the support of the aliquot assayed were the same as the composite sample collected, which is close to describing the stone to be removed by the truck load, the concentration reported would have met the Land Disposal Restriction criterion. Indeed, after the expenditure of additional mobilization, sampling and analytical costs, this was shown to be the case.

These expenditures could have been avoided by paying more attention to whether the support of the sample assayed was the same as the support required for making the desired decision. This requires that thoughtful, statistical consideration be given all aspects of sampling and subsampling with appropriate modification to "standard" protocols made as required.

In the present example, the sampling design should have specified that samples of stone of the size fraction to be removed be collected. Following Gy's theory (Gy, 1992; Pitard, 1993), the stone of the collected sample should have been crushed and mixed prior to selection of the aliquot for assay. Alternatively, solvent extraction could have been performed on the entire "as-collected" sample with subsampling of the "extractate."

What about Soil?

The problems associated with the sampling and assay of the stones are obvious because they are highly visual. Less visual are the similar inferential problems associated with the sampling and assay of all bulk materials. This is particularly true of soil. It is largely a matter of scale. One can easily observe the differences in size and composition of stone chips, but differences in the types and sizes of soil particles are less obvious to the eye of the sample collector.

Yet, because these differences are obvious to the assaying techniques, one must be extremely cautious in assuming the support of any analytical result. Care must be exercised in the sampling design, collection, and assay that the sampling-assaying processes do not contradict either the needs of the remediator or the dictates of the media and site correlation structure.

In situ soil is likely to exhibit a large degree of heterogeneity. Changes in soil type and moisture content may be extremely important to determinations of bio-availability of import to risk based decisions (for instance, see Miller and Zepp, 1987; Marple et al., 1987; and Umbreit et al., 1987). Consideration of such issues is absolutely essential if appropriate sampling designs are to be employed for making decisions regarding a meaningful observational unit.

A soil sample typically is sent to the analytical laboratory in a container that can be described as a "quart" jar. The contents of this container weigh approximately one kilogram depending, of course, on the soil moisture content and density. An aliquot is extracted from this container for assay by the laboratory according to the accepted assay protocol. The weight of the aliquot is 30 grams for organics and five (5) grams for metals (see Figure 1.2). Assuming an organic assay, there are 33 possible aliquots represented in the typical sampling container. Obviously, there are six times as many represented for a metals analysis.

Figure 1.2 Contrast between 30-gm Analytical Aliquot and 1-kg Field Sample

If an organics assay is to be performed, the organics are extracted with a solvent and the "extractate" concentrated to a volume of 10 milliliters. Approximately one-to-five *microliters* (about nine drops) are then taken from the 10 *milliliters* of "extractate" and injected into the gas chromatograph-mass spectrometer for analysis. Thus, there are approximated 2,000 possible injection volumes in the 10 milliliters of "extractate." This means that there are 66,000 possible measurements that can be made from a "quart" sample container. While assuming a certain lack of heterogeneity within a 10-milliliter volume of "extractate" may be reasonable, it may be yet another matter to assume a lack of heterogeneity among the 30-gram aliquots from the sample container (see Pitard, 1993).

A properly formed sample retains the heterogeneity of the entity sampled although, if thoroughly mixed, it may alter the distributional properties of the in situ material. However, the effects of gravity may well cause particle size segregation during transport. If the laboratory then takes the "first" 30-gram aliquot from the sample container, without thorough remixing of all the container's contents, the measurement provided by the assay cannot be assumed to be a reasonable estimate of the average concentration of the one kilogram sample.

New analytical techniques promise to exacerbate the problems of the support of the aliquot assayed. SW-846 Method 3051 is an approved analytical method for metals that requires a sample of less than 0.1 gram for microwave digestion. Methods currently pending approval employing autoextractors for organic analytes require less than 10 grams instead of the 30-gram aliquot used for Method 3500.

Assessment of Measurement Variation

How well a single assay result describes the average concentration desired can only be assessed by investigating the measurement variation. Unfortunately, such an assessment is usually only considered germane to the quality control/quality assurance portion of environmental investigations. Typically there is a requirement to have the analytical laboratory perform a duplicate analysis once every 20 samples. Duplicate analyses involve the selection of a second aliquot (subsample) from the submitted sample, and the preparation and analysis of it as if it were another sample. The results are usually reported in terms of the relative percent difference (RPD) between the two measurement results. This provides some measure of precision that not only includes the laboratory's ability to perform a measurement, but also the heterogeneity of the sample itself.

The RPD provides some estimate of the ability of an analytical measurement to characterize the material within the sample container. One often wonders what the result would be if a third, and perhaps a fourth aliquot were taken from the sample container and measured. The RPD, while meaningful to chemists, is not adequate to characterize the variation among measures on more than two aliquots from the same sample container. Therefore, more traditional statistical measures of precision are required, such as the variance or standard deviation.

In regard to determining the precision of the measurement, most everyone would agree that the 2,000 possible injections to the gas chromatograph/mass spectrometer from the 10 ml extractate would be expected to show a lack of heterogeneity. However, everyone might not agree that the 33 possible 30-gram aliquots within a sample container would also be lacking in heterogeneity.

Extending the sampling frame to "small" increments of time or space, introduces into the measurement system sources of possible heterogeneity that include the act of composite sample collection as well as those inherent to the media sampled. Gy (1992), Liggett (1995a, 1995b, 1995c), and Pitard (1993) provide excellent discussions of the statistical issues.

Having an adequate characterization of the measurement system variation may well assist in defining appropriate sampling designs for estimation of the desired average characteristic for the decision unit. Consider this example extracted from data contained in the site Remedial Investigation/Feasibility Study (RI/FS) reports for a confidential client. Similar data may be extracted from the RI/FS reports for almost any site.

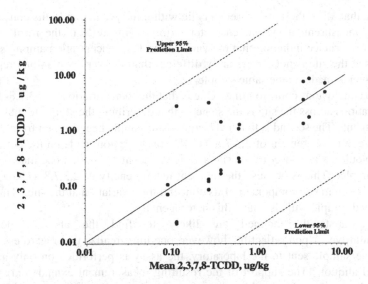

**Figure 1.3   Example Site 2,3,7,8-TCDD,
Sample Repeated Analyses versus Mean**

Figure 1.3 presents the results of duplicate measurements of 2,3,7,8-TCDD in soil samples taken at a particular site. These results are those reported in the quality assurance section of the site characterization report and are plotted against their respective means. The "prediction limits" shown in this figure will, with 95 percent confidence, contain an additional single measurement (Hahn 1970a, 1970b). If one considers all the measurements of 2,3,7,8-TCDD made at the site and plots them versus their mean, the result is shown in Figure 1.4.

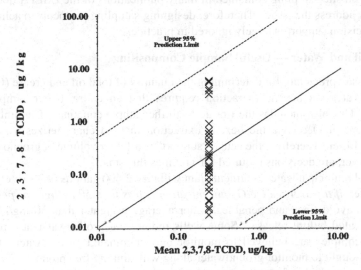

**Figure 1.4   Example Site 2,3,7,8-TCDD, All Site Samples
versus Their Mean**

Note that all of these measurements lie within the prediction limits constructed from the measurement system characterization. This reflects the results of an analysis of variance indicting that the variation in log-concentration among sample locations at the site is not significantly different than the variation among repeated measurements made on the same sample.

Two conclusions come to mind. One is that the total variation of 2,3,7,8-TCDD concentrations across the site is the same as that describing the ability to make such measurement. The second is that had a composite sample been formed from the soil at this site, a measurement of 2,3,7,8-TCDD concentration made on the composite sample would be no closer to the site average concentration than one made on any single sample. This is because the inherent heterogeneity of 2,3,7,8-TCDD in the soil matrix is a major component of its concentration variation at the site. Thus, the composited sample will also have this heterogeneity.

The statistically inclined are likely to find the above conclusion counterintuitive. Upon reflection, however, one must realize that regardless of the size of the sample sent to the laboratory, the assay is performed on only a small fractional aliquot. The support of the resulting measurement extends only to the assayed aliquot. In order to achieve support equivalent to the size of the sample sent, it is necessary to either increase the physical size of the aliquot assayed, or increase the number of aliquots assayed per sample and average their results. Alternatively, one could grind and homogenize the entire sample sent before taking the aliquot for assay. In light of this, one wonders what is really implied in basing a risk assessment for 2,3,7,8-TCDD on the upper 95 percent confidence limit for the mean concentration of 30-gram aliquots of soil.

In other words, more thought should be given to the support associated with an analytical result during sampling design. Unfortunately, historically the "relevant guidance" on site sampling contained in many publications of the USEPA does not adequately address the issue. Therefore, designing sampling protocols to achieve a desired decision support is largely ignored in practice.

Mixing Oil and Water — Useful Sample Compositing

The assay procedure for determining the quantity of total oil and grease (O&G) in groundwater via hexane extraction requires that an entire 1-liter sample be extracted. This also includes the rinsate from the sample container. Certainly, the measurement of O&G via the hexane extraction method characterizes a sample volume of 1 liter. Therefore, the actual "support" is a 1-liter volume of groundwater. Rarely, if ever, are decisions required for volumes this small.

A local municipal water treatment plant will take 2,400 gallons (9,085 liters) per day of water, *if the average O&G concentration is less than 50 milligrams per liter (mg/l)*. To avoid fines and penalties, water averaging greater than 50 mg/l O&G must be treated before release. Some wells monitoring groundwater at a former industrial complex are believed to monitor uncontaminated groundwater. Other wells are thought to monitor groundwater along with sinking free product. The task is to develop a means of monitoring groundwater to be sent to the local municipal treatment plant.

Figure 1.5 presents the results of a sampling program designed to estimate the variation of O&G measurements with 1-liter support. This program involved the repeated collection of 1-liter grab samples of groundwater from the various monitoring wells at the site over a period of several hours. Obviously, a single grab sample measurement for O&G does not provide adequate support for decisions regarding the average O&G concentration of 2,400 gallons of groundwater. However, being able to estimate the within-well mean square assists the development of an appropriate sampling design for monitoring discharged groundwater.

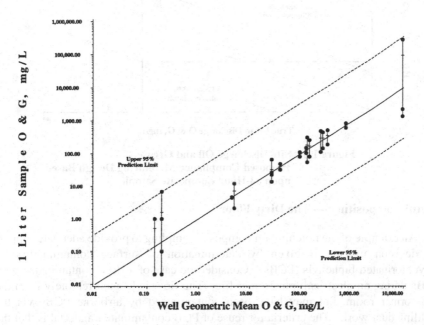

Figure 1.5 **Groundwater Oil and Grease — Hexane Extraction, Individual 1-Liter Sample Analyses by Source Well Geometric Mean**

Confidence limits for the true mean O&G concentration as would be estimated from composite samples having 24-hour support are presented in Figure 1.6. This certainly suggests that an assay of a flow-weighted composite sample would provide a reasonable estimate of the true mean O&G concentration during some interesting time span.

The exercise also provides material to begin drafting discharge permit conditions based upon a composite over a 24-hour period. These might be stated as follows: (1) If the assay of the composite sample is less than 24 mg/l O&G, then the discharge criteria is met. (2) If this assay result is greater than 102 mg/l, then the discharge criteria has not been met. While this example may seem intuitively obvious to statisticians, it is this author's experience that the concept is totally foreign to many engineers and environmental managers.

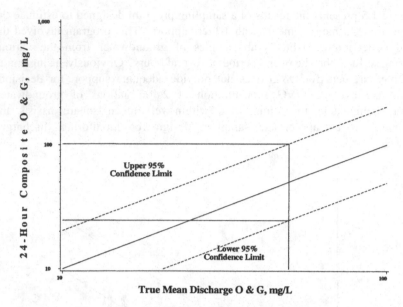

Figure 1.6   Site Discharge Oil and Grease,
Proposed Compliance Monitoring Design Based
upon 24-Hour Composite Sample

Useful Compositing — The Dirty Floor

An example of the potential for composite sampling to provide adequate support for decision making is given by determination of surface contamination by polychlorinated biphenyls (PCBs). Consider the case of a floor contaminated with PCBs during an electrical transformer fire. The floor is located remotely from the transformer room, but may have been contaminated by airborne PCBs via the building duct work. The criteria for reuse of PCB contaminated material is that the PCB concentration must be less than 10 micrograms per 100 square centimeters (μg/100 cm^2). That is, the entire surface must have a surface concentration of less than 10 μg/100 cm^2.

The determination of surface contamination is usually via "wipe" sampling. Here a treated filter type material is used to wipe the surface using a template that restricts the amount of surface wiped to 100 cm^2. The "wipes" are packaged individually and sent to the laboratory for extraction and assay. The final chemical measurement is preformed on an aliquot of the "extractate."

Suppose that the floor has been appropriately sampled (Ubinger 1987). A determination regarding the "cleanliness" of the floor may be made from an assay of composited extractate if the following conditions are satisfied. One, the detection limit of the analytical method must be at least the same fraction of the criteria as the number of samples composited. In other words, if the extractate from four wipe samples is to be composited, the method detection limit must be 2.5 μg/100 cm^2 or less. Two, it must be assumed that the aliquot taken from the sample extractate for composite formation is "representative" of the entity from which it was taken. This assumes that the wipe sample extractate lacks heterogeneity when the subsample aliquot is selected.

If the assay results are less than 2.5 μg/100 cm^2, then the floor will be declared clean and appropriate for reuse. If, on the other hand, the result is greater than 2.5 μg/100 cm^2, the remaining extractate from each individual sample may be assayed to determine if the floor is uniformly contaminated, or if only a portion of it exceeds 10 μg/100 cm^2.

Comments on Stuff Blowing in the Wind

Air quality measurements are inherently made on samples composited over time. Most are weighted by the air flow rate through the sampling device. The only air quality measure that comes to mind as not being a flow-weighted composite is a particulate deposition measurement. It appears to this writer that it is the usual interpretation that air quality measurements made by a specific monitor represent the quality of ambient air in the general region of the monitor. It also appears to this writer that it is legitimate to ask how large an ambient air region is described by such a measurement.

Figure 1.7 illustrates the differences in hourly particulate (PM$_{10}$) concentrations between co-located monitors. Figure 1.8 illustrates the differences in hourly PM$_{10}$ between two monitors separated by approximately 10 feet. All of these monitors were located at the Lincoln Monitoring Site in Allegheny County, Pennsylvania. This is an industrial area with a multiplicity of potential sources of PM$_{10}$. The inlets for the co-located monitors are at essentially the same location.

The observed differences in hourly PM$_{10}$ measurements for the monitors with 10-foot separation is interesting for several reasons. The large magnitude of some of these differences certainly will affect the difference in the 24-hour average concentrations. This magnitude is as much as 70–100 μg/cubic meter on June 17 and 19. During periods when the measured concentration is near the 150-μg/cubic meter standard, such a difference could affect the determination of attainment. Because the standard is health based and presumes a 24-hour average exposure, the support of the ambient air quality measurement takes on increased importance.

If the support of an ambient air quality measurement is only in regard to inferences regarding a rather small volume of air, say within a 10-foot semisphere around the monitor, it is unlikely to describe the exposure of anyone not at the monitor site. Certainly, there is no support from this composite sample measurement for the making of inferences regarding air quality within a large region unless it can be demonstrated that there is no heterogeneity within the region. This requires a study of the measurement system variation utilizing monitors placed at varying distances apart. In truth, any ambient air quality monitor can only composite a sample of air precisely impinging on the monitor's inlet. It cannot form an adequate composite sample of air in any reasonable spatial region surrounding that monitor.

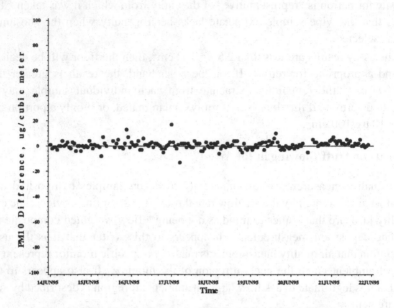

**Figure 1.7   Hourly Particulate (PM$_{10}$) Monitoring Results,
Single Monitoring Site, June 14–21, 1995,
Differences between Co-located Monitoring Devices**

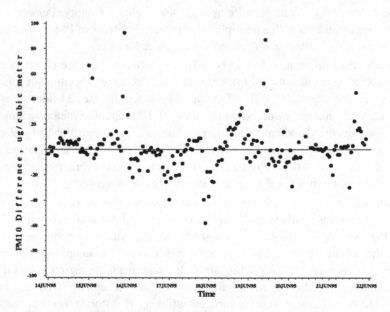

**Figure 1.8   Hourly Particulate (PM$_{10}$) Monitoring Results,
Single Monitoring Site, June 14–21, 1995,
Differences between Monitoring Devices 10 Feet Apart**

A Note on Composite Sampling

The previous examples deal largely with sample collection schemes involving the combination of logically smaller physical entities collected over time and/or space. Considering Gy's sampling theory, one might argue that all environmental samples are "composite" samples.

It should be intuitive that a decision regarding the average contaminant concentration over one-half an acre could not be well made from a single-kilogram sample of soil taken at a randomly chosen location within the plot. Obviously, a much more sound decision-making basis is to average the contaminant concentration results from a number of 1-kilogram samples taken from the plot. If the formation of a composite sample can be thought of as the "mechanical averaging" of concentration, then composite sampling appears to provide for great efficiency in cost-effective decision making. This of course assumes that the formation of the composite sample and its assay truly retain the "support" intended by the sampling design. The examples above have shown that unless care is used in the sample formation and analyses, the desired decision support may not be achieved.

Webster's (1987) defines composite as (1) made up of distinct parts, and (2) combining the *typical or essential characteristics* of individuals making up a group. Pitard (1993, p. 10) defines a composite sample as a "sample made up of the reunion of several distinct subsamples." These definitions certainly describe an entity that should retain the "average" properties of the whole consonant with the notion of support.

On the surface, composite sampling has a great deal of appeal. In practice this appeal is largely economic in that there is a promise of decreased sample processing, shipping, and assay cost. However, if one is not very careful, this economy may come at a large cost due to incorrect decision making. While the desired support may be carefully built into the formation of a composite soil sample, it may be poorly reflected in the final assay result.

This is certainly a problem that can be corrected by appropriate design. However, the statistician frequently is consulted only as a last resort. In such instances, we find ourselves practicing statistics in retrospection. Here the statistician needs to be particularly attuned to precisely defining the support of the measurement made before assisting with any inference. Failure to do so would just exacerbate the confusion as discussed by Englund and Heravi (1994).

Sampling Design

Systematic planning for sample collection has been required by USEPA executive order since 1984 (USEPA, 1998). Based upon the author's experience, much of the required planning effort is focused on the minute details of sample collection, preservation, shipping, and analysis. Forgotten are seeking answers to the following three very important questions:

- What does one really wish to know?
- What does one already know?
- How certain does one wish to be about the result?

These are questions that statisticians ask at the very beginning of any sampling program design. They are invited as soon as the statistician hears, "How many samples do I need to take?" All too often it is not the answers to these questions that turn out to be important to decision making, but the process of seeking them. Frequently the statistician finds that the problem has not been very well defined and his asking of pointed questions gives focus to the *real* purpose for sample collection. William Lurie nicely described this phenomenon in 1958 in his classic article, "The Impertinent Questioner: The Scientist's Guide to the Statistician's Mind."

Many of the examples in this chapter illustrate what happens when the process of seeking the definition for sample collection is short circuited or ignored. The result is lack of ability to make the desired decision, increased costs of resampling and analysis, and unnecessary delays in environmental decision making. The process of defining the desired sample collection protocol is very much an interactive and iterative one. An outline of this process is nicely provided by the USEPA's Data Quality Objectives (DQO) Process.

Figure 1.9 provides a schematic diagram of the DQO process. Detailed discussion of the process can be found in the appropriate USEPA guidance (USEPA, 1994a). Note that the number and placement of the actual samples is not accomplished until Step 7 of the DQO process. Most of the effort in designing a sampling plan is, or should be, expended in Steps 1 through 5. An applied statistician, schooled in the art of asking the right questions, can greatly assist in optimizing this effort (as described by Lurie, 1958).

The applied statistician is also skilled in deciding which of the widely published formulae and approaches to the design of environmental sampling schemes truly satisfy the site specific assumptions uncovered during Steps 1–6. (See Gilbert, 1987; USEPA, 1986, 1989, 1994b, 1996a, and 1996b.) Failure to adequately follow this process only results in the generation of data that do not impact on the desired decision as indicated by several of the examples at the beginning of this chapter.

Step 8 of the process, EVALUATE, is only tacitly discussed in the referenced USEPA guidance. Careful review of all aspects of the sampling design before implementation has the potential for a great deal of savings in resampling and reanalysis costs. This is evident in the "Story of the Stones" discussed at the beginning of this chapter. Had someone critically evaluated the initial design before going into the field, they would have realized that instructions to the laboratory should have specifically indicated the extraction of all stones collected.

Evaluation will often trigger one or more iterations through the DQO process. Sampling design is very much a process of interaction among statistician, decision maker, and field and laboratory personnel. This interaction frequently involves compromise and sometimes redefinition of the problem. Only after everyone is convinced that the actual support of the samples to be collected will be adequate to make the decisions desired, should we head to the field.

Institutional Impediments to Sampling Design

In the authors' opinion, there is a major impediment to the DQO process and adequate environmental sampling design. This is the time honored practice of

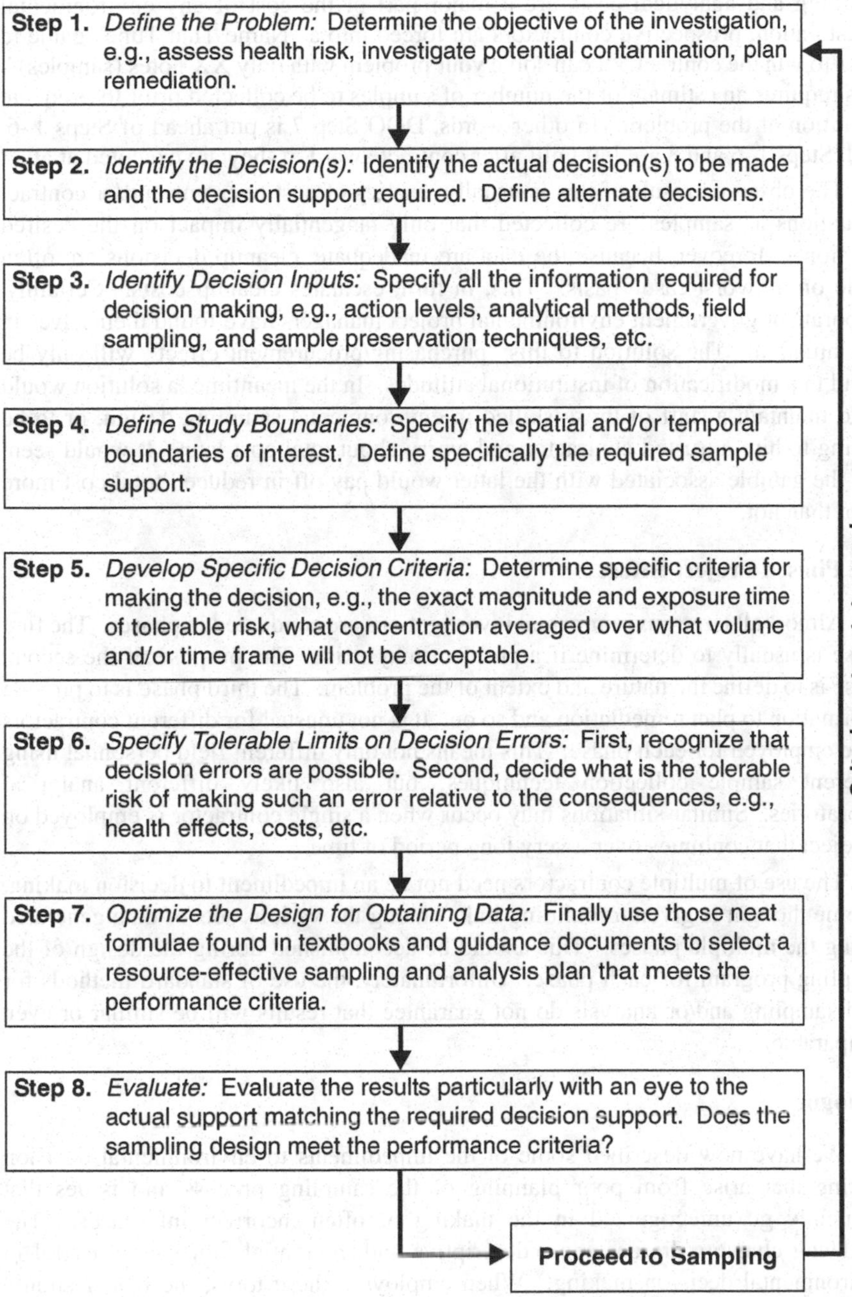

Step 1. *Define the Problem:* Determine the objective of the investigation, e.g., assess health risk, investigate potential contamination, plan remediation.

Step 2. *Identify the Decision(s):* Identify the actual decision(s) to be made and the decision support required. Define alternate decisions.

Step 3. *Identify Decision Inputs:* Specify all the information required for decision making, e.g., action levels, analytical methods, field sampling, and sample preservation techniques, etc.

Step 4. *Define Study Boundaries:* Specify the spatial and/or temporal boundaries of interest. Define specifically the required sample support.

Step 5. *Develop Specific Decision Criteria:* Determine specific criteria for making the decision, e.g., the exact magnitude and exposure time of tolerable risk, what concentration averaged over what volume and/or time frame will not be acceptable.

Step 6. *Specify Tolerable Limits on Decision Errors:* First, recognize that decision errors are possible. Second, decide what is the tolerable risk of making such an error relative to the consequences, e.g., health effects, costs, etc.

Step 7. *Optimize the Design for Obtaining Data:* Finally use those neat formulae found in textbooks and guidance documents to select a resource-effective sampling and analysis plan that meets the performance criteria.

Step 8. *Evaluate:* Evaluate the results particularly with an eye to the actual support matching the required decision support. Does the sampling design meet the performance criteria?

Criteria not met; try again

Proceed to Sampling

Figure 1.9 The Data Quality Objectives Process

accepting the lowest proposed "cost" of an environmental investigation. Since the sampling and analytical costs are a major part of the cost of any environmental investigation, prospective contractors are forced into a "Name That Tune" game in order to win the contract. "I can solve your problem with only XX notes (samples)." This requires an estimate of the number of samples to be collected prior to adequate definition of the problem. In other words, DQO Step 7 is put ahead of Steps 1–6. And, Steps 1–6 and 8 are left until after contract award, if they are executed at all.

The observed result of this is usually a series of cost overruns and/or contract escalations as samples are collected that only tangentially impact on the desired decision. Moreover, because the data are inadequate, cleanup decisions are often made on a "worst-case" basis. This, in turn, escalates cleanup costs. Certainly, corporate or government environmental project managers have found themselves in this situation. The solution to this "purchasing/procurement effect" will only be found in a modification of institutional attitudes. In the meantime, a solution would be to maintain a staff of those skilled in environmental sampling design, or to be willing to hire a trusted contractor and worry about total cost later. It would seem that the gamble associated with the latter would pay off in reduced total cost more often than not.

The Phased Project Effect

Almost all large environmental investigations are conducted in phases. The first phase is usually to determine if a problem may exist. The purpose of the second phase is to define the nature and extent of the problem. The third phase is to provide information to plan remediation and so on. It is not unusual for different contractors to be employed for each phase. This means not only different field personnel using different sample collection techniques, but also likely different analytical laboratories. Similar situations may occur when a single contractor is employed on a project that continues over a very long period of time.

The use of multiple contractors need not be an impediment to decision making, if some thought is given to building links among the various sets of data generated during the multiple phases. This should be accomplished during the design of the sampling program for each phase. Unfortunately, the use of standard methods for field sampling and/or analysis do not guarantee that results will be similar or even comparable.

Epilogue

We have now described some of the impediments to environmental decision making that arise from poor planning of the sampling process and issues that frequently go unrecognized in the making of often incorrect inferences. The following chapters discuss some descriptive and inferential tools found useful in environmental decision making. When employing these tools, the reader should always ask whether the resulting statistic has the appropriate support for the decision that is desired.

References

Englund, E. J. and Heravi, N., 1994, "Phased Sampling for Soil Remediation," *Environmental and Ecological Statistics*, 1: 247–263.

Flatman, G. T. and Yfantis, A. A., 1996, "Geostatistical Sampling Designs for Hazardous Waste Site," *Principles of Environmental Sampling*, ed. L. Keith, American Chemical Society, pp. 779–801.

Gilbert, R. O., 1987, *Statistical Methods for Environmental Pollution Monitoring*, Van Nostrand Reinhold, New York.

Gy, P. M., 1992, *Sampling of Heterogeneous and Dynamic Material Systems: Theories of Heterogeneity, Sampling, and Homogenizing*, Elsevier, Amsterdam.

Hahn, G. J., 1970a, "Statistical Intervals for a Normal Population, Part I. Tables, Examples and Applications," *Journal of Quality Technology*, 2: 115–125.

Hahn, G. J., 1970b, "Statistical Intervals for a Normal Population, Part II. Formulas, Assumptions, Some Derivations," *Journal of Quality Technology*, 2: 195-206.

Liggett, W. S., and Inn, K. G. W., 1995a, "Pilot Studies for Improving Sampling Protocols," *Principles of Environmental Sampling*, ed. L. Keith, American Chemical Society, Washington, D.C.

Liggett, W. S., 1995b, "Functional Errors-in-Variables Models in Measurement Optimization Experiments," *1994 Proceedings of the Section on Physical and Engineering Sciences*, American Statistical Association, Alexandria, VA.

Liggett, W. S., 1995c, "Right Measurement Tools in the Reinvention of EPA," *Corporate Environmental Strategy*, 3: 75–78.

Luric, William, 1958, "The Impertinent Questioner: The Scientist's Guide to the Statistician's Mind," *American Scientist*, March.

Marple, L., Brunck, R., Berridge, B., and Throop, L., 1987, "Experimental and Calculated Physical Constants for 2,3,7,8-Tetrachlorodibenzo-*p*-dioxin," *Solving Hazardous Waste Problems Learning from Dioxins*, ed. J. Exner, American Chemical Society, Washington, D.C., pp. 105–113.

Miller, G. C. and Zepp, R. G., 1987, "2,3,7,8-Tetrachlorodibenzo-*p*-dioxin: Environmental Chemistry," *Solving Hazardous Waste Problems Learning from Dioxins*, ed. J. Exner, American Chemical Society, Washington, D.C., pp. 82–93.

Neptune, D., Brantly, E. P., Messner, M. J., and Michael, D. I., 1990, "Quantitative Decision Making in Superfund: A Data Quality Objectives Case Study," *Hazardous Material Control*, May/June.

Olea, R., 1991, *Geostatistical Glossary and Multilingual Dictionary*, Oxford University Press, New York.

Pitard, F. F., 1993, *Pierre Gy's Sampling Theory and Sampling Practice, Second Edition*, CRC Press, Boca Raton, FL.

Ubinger, E. B., 1987, "Statistically Valid Sampling Strategies for PCB Contamination," *Presented at the EPRI Seminar on PCB Contamination*, Kansas City, MO, October 6–9.

Umbreit, T. H., Hesse, E. J., and Gallo, M. A., 1987, "Differential Bioavailability of 2,3,7,8-Tetrachlorodibenzo-*p*-dioxin from Contaminated Soils," *Solving Hazardous Waste Problems Learning from Dioxins*, ed. J. Exner, American Chemical Society, Washington, D.C., pp. 131–139.

USEPA, 1986, *Test Methods for Evaluating Solid waste (SW-846): Physical/ Chemical Methods*, Third Edition, Office of Solid Waste.

USEPA, 1989, *Risk Assessment Guidance for Superfund: Human Health Evaluation Manual Part A*, EPA/540/1-89/002.

USEPA, 1994a, *Guidance for the Data Quality Objectives Process, EPA QA/G-4.*

USEPA, 1994b, *Data Quality Objectives Decision Error Feasibility Trials (DQO/DEFT), User's Guide*, Version 4, EPA QA/G-4D.

USEPA, 1996a, *Soil Screening Guidance: Technical Background Document*, EPA/540/R95/128.

USEPA, 1996b, *Soil Screening Guidance: User's Guide*, Pub. 9355.4-23.

USEPA, 1998, *EPA Order 5360.1, Policy and Program Requirements for the Mandatory Agency-Wide Quality System.*

Webster's, 1987, *Webster's Ninth New Collegiate Dictionary*, Merriam-Webster Inc., Springfield, MA.

CHAPTER 2

Basic Tools and Concepts
Description of Data

> The goal of statistics is to gain information from data. The first step is to display the data in a graph so that our eyes can take in the overall pattern and spot unusual observations. Next, we often summarize specific aspects of the data, such as the average of a value, by numerical measures. As we study graphs and numerical summaries, we keep firmly in mind where the data come from and what we hope to learn from them. Graphs and numbers are not ends in themselves, but aids to understanding. (Moore and McCabe, 1993)

Every study begins with a sample, or a set of measurements, which is "representative" in some sense, of some population of possible measurements. For example, if we are concerned with PCB contamination of surfaces in a building where a transformer fire has occurred, our sample might be a set of 20 surface wipe samples chosen to represent the population of possible surface contamination measurements. Similarly, if we are interested in the level of pesticide present in individual apples, our sample might be a set of 50 apples chosen to be representative of all apples (or perhaps all apples treated with pesticide). Our focus here is the set of statistical tools one can use to describe a sample, and the use of these sample statistics to infer the characteristics of the underlying population of measurements.

Central Tendency or Location

The Arithmetic Mean

Perhaps the first question one asks about a sample is what is a typical value for the sample. Usually this is answered by calculating a value that is in the middle of the sample measurements. Here we have a number of choices. We can calculate the arithmetic mean, $\bar{x}$, whose value is given by:

$$\bar{x} = \frac{\Sigma x_i}{N} \qquad [2.1]$$

where the x_i's are the individual sample measurements and N is the sample size.

The Geometric Mean

Alternatively, we can calculate the geometric mean, GM_x, given by:

$$GM(x) = \exp\Sigma \ln(x_i) / N \qquad [2.2]$$

19

That is, GM(x) is the antilogarithm of the mean of the logarithms of the data value. Note that for the GM to be defined, all x's must be greater than zero.

If we calculate ln (GM(x)), this is called the logarithmic mean, LM(x), and is simply the arithmetic mean of the log-transformed x's.

The Median

The median, M, is another estimator of central tendency. It is given by the 50th percentile of the data. If we have a sample of size N, sorted from smallest to largest (e.g., x_1 is the smallest observation and x_N is the largest) and N is odd, the median is given by x_j. Here j is given as:

$$j = ((N - 1) / 2) + 1 \qquad [2.3]$$

That is, if we have 11 observations the median is equal to the 6th largest and if we have 7 observations, the median is equal to the 4th largest. When N is an even number, the median is given as:

$$M = (x_j + x_k) / 2 \qquad [2.4]$$

In Equation [2.4], j and k are equal to (N/2) and ((N/2) + 1), respectively. For example if we had 12 observations, the median would equal the average of the 6th and 7th largest observations. If we had 22 observations, the median would equal the average of the 11th and 12th largest values.

Discussion

While there are other values, such as the mode of the data (the most frequent value) or the harmonic mean (the reciprocal of the mean of the 1/x values), the arithmetic mean, the geometric mean and the median are the three measures of central tendency routinely used in environmental quality investigations. The logarithmic mean is not of interest as a measure of central tendency because it is in transformed units (ln (concentration)), but does arise in considerations of hypothesis tests.

Note also that all of these measures of sample central tendency are expected to represent the corresponding quantities in the population (often termed the "parent" population) from which the sample was drawn. That is, as the sample size becomes large, the difference between, for example, $\bar{x}$ and μ (the parametric or "true" arithmetic mean) becomes smaller and smaller, and in the limit is zero. In statistical terms these "sample statistics" are unbiased estimators of the corresponding population parameters.

Dispersion

By dispersion we mean how spread out the data are. For example, say we have two areas, both with a median concentration of 5 ppm for some compound of interest. However, in the first area the 95th percentile concentration is 25 ppm while in the second, the 95th percentile concentration is 100 ppm. One might argue that the central tendency or location of the compound of interest is similar in these areas (or not, depending on the purpose of our investigation; see Chapter 3), but the second area clearly has a much greater spread or dispersion of concentrations than the first. The question is, how can this difference be expressed?

The Sample Range

One possibility is the sample range, W, which is given by:

$$W = x_{max} - x_{min} \qquad [2.5]$$

that is, W is the difference between the largest and smallest sample values. This is certainly a good measure of the dispersion of the sample, but is less useful in describing the underlying population. The reason that this is not too useful as a description of the population dispersion is that its magnitude is a function of both the actual dispersion of the population and the size of the sample. We can show this as follows:

1. The median percentile, mp_{max}, of the population that the largest value in a sample of N observations will represent is given by:

$$mp_{max} = 0.5^{1/N}$$

   that is, if we have a sample of 10 observations, mp_{max} equals $0.5^{1/10}$ or 0.933. If instead we have a sample of 50 observations, mp_{max} equals $0.5^{1/50}$ or 0.986. That is, if the sample size is 10, the largest value in the sample will have a 50-50 chance of being above or below the 93.3rd percentile of the population from which the sample was drawn. However, if the sample size is 50, the largest value in the sample will have a 50-50 chance of being above or below the 98.6th percentile of the population from which the sample was drawn.

2. The median percentile, mp_{min}, of the population that the smallest value in a sample of N observations will represent is given by:

$$mp_{min} = 1 - 0.5^{1/N}$$

   For a sample of 10 observations, mp_{min} equals or 0.0.067, and for a sample of 50 observations, mp_{min} equals 0.0.014.

3. Thus for a sample of 10 the range will tend to be the difference between the 6.7th and 93.3rd percentiles of the population from which the sample was drawn, while for a sample of 50, the range will tend to be the difference between the 1.4th and 98.6th percentiles of the population from which the sample was drawn. More generally, as the sample becomes larger and larger, the range represents the difference between more and more extreme high and low percentiles of the population.

This is why the sample range is a function of both the dispersion of the population and the sample size. For equal sample sizes the range will tend to be larger for a population with greater dispersion, but for populations with the same dispersion the sample range will larger for larger N.

The Interquartile Range

One way to fix the problem of the range depending on the sample size is to calculate the difference between fixed percentiles of the data. The first problem encountered is the calculation of percentiles. We will use the following procedure:

1. Sort the N sample observations from smallest to largest.

2. Let the rank of an observation be I, its list index value. That is, the smallest observation has rank 1, the second smallest has rank 2, and so on, up to the largest value that has rank N.

3. The cumulative probability, P_I, of rank I is given by:

$$P_I = (I - 3/8) / (N + 1/4) \qquad [2.6]$$

This cumulative probability calculation gives excellent agreement with median probability calculated from the theory of order statistics. (Looney and Gulledge, 1995)

To get values for cumulative probabilities not associated with a given rank.

1. Pick the cumulative probability, CP, of interest (e.g., 0.75).

2. Pick the P_I value of the rank just less than CP. The next rank has cumulative probability value P_{I+1} (note that one cannot calculate a value for cumulative probabilities less than P_1 or greater than P_N).

3. Let the values associated with these ranks be given by $V_I = V_L$ and $V_{I+1} = V_U$.

4. Now if we assume probability is uniform between $P_I = P_L$ and $P_{I+1} = P_U$ it is true that:

$$(CP - P_L) / (P_U - P_L) = (V_{CP} - V_L) / (V_U - V_L) \qquad [2.7]$$

where V_{CP} is the CP (e.g., 0.75) cumulative probability, V_L is the value associated with the lower end of the probability interval, P_L and V_U is the value associated with the upper end of the probability interval, P_U. One can rearrange [2.6] to obtain $V_{0.75}$ as follows:

$$V_{0.75} = ((V_U - V_L) \times (0.75 - P_L) / (P_U - P_L)) + V_L \qquad [2.8]$$

This is general for all cumulative probabilities that we can calculate. Note that one cannot calculate a value for cumulative probabilities less than P_1 or greater than P_N because in the first case P_L is undefined and in the second P_U is undefined. That is, if we wish to calculate the value associated with a cumulative probability of 0.95 in a sample of 10 observations, we find that we cannot because P_{10} is only about 0.94.

As one might expect from the title of this section, the interquartile range, IQ, given by:

$$IQ = V_{0.75} - V_{0.25} \qquad [2.9]$$

is a commonly used measure of dispersion. It has the advantage that its expected width does not vary with sample size and is defined (calculable) for samples as small as 3.

The Variance and Standard Deviation

The sample variance, S^2 is defined as:

$$S^2 = \frac{\Sigma(x_i - \bar{x})^2}{(N-1)} \qquad [2.10]$$

where the x_i's are the individual sample measurements and N is the sample size. Note that one sometimes also sees the formula:

$$\sigma^2 = \frac{\Sigma(x_i - \bar{x})^2}{(N)} \qquad [2.11]$$

Here σ^2 is the *population* variance. The difference between [2.10] and [2.11] is the denominator. The $(N-1)$ term is used in [2.10] because using N as in [2.11] with any finite sample will result in an estimate of S^2, which is too small relative to the true value of σ^2. Equation [2.11] is offered as an option in some spreadsheet programs, and is sometimes mistakenly used in the calculation of sample statistics. This is always wrong. One should always use [2.10] with sample data because it always gives a more accurate estimate of the true σ^2 value.

The sample standard deviation, S is given by:

$$S = (S^2)^{1/2} \qquad [2.12]$$

that is, the sample standard deviation is the square root of the sample variance.

It is easy to see that S and S^2 reflect the dispersion of the measurements. The variance is, for large samples, approximately equal to the average squared deviation of the observations from the sample mean, which as the observations get more and more spread out, will get larger and larger.

If we can assume that the observations follow a normal distribution, we can also use $\bar{x}$ and s to calculate estimates of extreme percentiles. We will consider this at some length in our discussion of the normal distribution.

The Logarithmic and Geometric Variance and Standard Deviation

Just as we can calculate the arithmetic mean of the log transformed observations, LM(x), and its anti-log, GM(x), we can also calculate the variance and standard deviation of these log-transformed measurements, termed the logarithmic variance, LV(y), and logarithmic standard deviation LSD(x), and their anti-logs, termed the geometric variance, GV(y), and geometric standard deviation, GSD(x), respectively. These measures of dispersion find application when the log-transformed measurements follow a normal distribution, which means that the measurements themselves follow what is termed a log-normal distribution.

The Coefficient of Variation (CV)

The sample CV is defined as:

$$CV = (S/\bar{x}) \bullet 100 \qquad\qquad [2.13]$$

that is, it is the standard deviation expressed as a percentage of the sample mean. Note that S and x have the same units. That is, if our measurements are in units of ppm, then both S and x are in ppm. Thus, the CV is always unitless. The CV is useful because it is a measure of relative variability. For example, if we have a measurement method for a compound, and have done ten replicates each at standard concentrations of 10 and 100 ppm, we might well be interested in relative rather than absolute precision because a 5% error at 10 ppm is 0.5 ppm, but the same relative error at 100 ppm is 5 ppm. Calculation of the CV would show that while the absolute dispersion at 100 ppm is much larger than that at 5 ppm, the relative dispersion of the two sets of measurements is equivalent.

Discussion

The proper measure of the dispersion of one's data depends on the question one wants to ask. The sample range does not estimate any parameter of the parent population, but it does give a very clear idea of the spread of the sample values. The interquartile range does estimate the population interquartile range and clearly shows the spread between the 25th and 75th percentiles. Moreover, this is the only dispersion estimate that we will discuss that accurately reflects the same dispersion measure of the parent population and that does not depend on any specific assumed distribution for its interpretation. The arithmetic variance and standard deviation are primarily important when the population follows a normal distribution, because these statistics can help us estimate error bounds and conduct hypothesis tests. The situation with the logarithmic and geometric variance and standard deviation is similar. These dispersion estimators are primarily important when the population follows a log-normal distribution.

Some Simple Plots

The preceding sections have discussed some basic measures of location (arithmetic mean, geometric mean, median) and dispersion (range, interquartile range, variance, and standard deviation). However, if one wants to get an idea of what the data "look like," perhaps the best approach is to plot the data (Tufte, 1983; Cleveland, 1993; Tukey, 1977). There are many options for plotting data to get an idea of its form, but we will discuss only three here.

Box and Whisker Plots

The first, called a "box and whisker plot" (Tukey, 1977), is shown in Figure 2.1. This plot is constructed using the median and the interquartile range (IQR). The IQR defines the height of the box, while the median is shown as a line within the box. The whiskers are drawn from the upper and lower hinges ((UH and LH; top and bottom of the box; 75th and 25th percentiles) to the largest and smallest observed values within 1.5 times the IQR of the UH and LH, respectively. Values between 1.5 and 3 times the IQR above or below the UH or LH are plotted as "*" and are termed

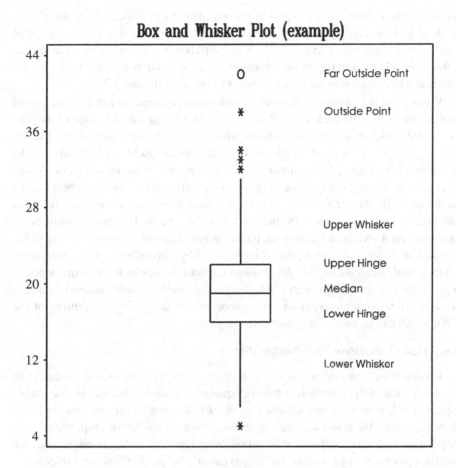

Figure 2.1 A Sample Box Plot

"outside points." Values beyond 3 times the IQR above or below the UH and LH values are plotted as "o" and are termed "far outside values." The value of this plot is that is conveys a great amount of information about the form of one's data in a very simple form. It shows central tendency and dispersion as well as whether there are any extremely large or small values. In addition one can assess whether the data are symmetric in the sense that values seem to be similarly dispersed above and below the median (see Figure 2.2D) or are "skewed" in the sense that there is a long tail toward high or low values (see Figure 2.4).

Dot Plots and Histograms

A dot plot (Figure 2.2A) is generated by sorting the data into "bins" of specified width (here about 0.2) and plotting the points in a bin as a stack of dots (hence the name dot plot). Such plots can give a general idea of the shape and spread of a set of data, and are very simple to interpret. Note also that the dot plot is similar in concept to a histogram (Figure 2.2B). A key difference is that when data are sparse, a dot plot will still provide useful information on the location and spread of the data whereas a histogram may be rather difficult to interpret (Figure 2.2B).

When there are substantial number of data points, histograms can provide a good look at the relative frequency distribution of x. In a histogram the range of the data is divided into a set of intervals of fixed width (e.g., if the data range from 1 to 10, we might pick an interval width of 1, which would yield 10 intervals). The histogram is constructed by counting up the data points whose value lies in a given interval and drawing a bar whose height corresponds to the number of observations in the interval. In practice the scale for the heights of the bars may be in either absolute or relative units. In the first case the scale is simply numbers of observations, k, while in the second, the scale is in relative frequency, which is the fraction of the total sample, N, that is represented by a given bar (relative frequency = k/N). Both views are useful. An absolute scale allows one to see how many points a given interval contains, which can be useful for small- to medium-sized data sets, while the relative scale provides information on the frequency distribution of the data, which can be particularly useful for large data sets.

Empirical Cumulative Distribution Plots

If we sort the observations in a sample from smallest to largest, we can calculate the proportion of the sample less than or equal to a given observation by the simple equation I/N, where N is the sample size and I is the rank of the observation in the sorted sample. We could also calculate the expected cumulative proportion of the population associated with the observation using Equation [2.6]. In either case, we can then plot the x's against their calculated cumulative proportions to produce a plot like that shown in Figure 2.2C. These empirical cumulative distribution plots can show how rapidly data values increase with increasing rank, and are also useful in determining what fraction of the observations are above some value of interest.

A. An Example Dot Plot

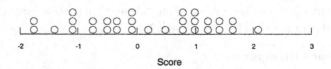

Score

Figure 2.2A Examples of Some Useful Plot Types

B. An Example Histogram

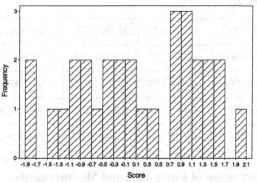

Figure 2.2B Examples of Some Useful Plot Types

C. An Example Empirical Cumulative Distribution Plot

Figure 2.2C Examples of Some Useful Plot Types

D. An Example Box and Whisker Plot

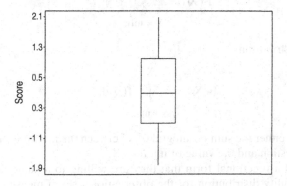

Figure 2.2D Examples of Some Useful Plot Types

28   *Basic Tools and Concepts Description of Data*

Table 2.1
Data Used in Figure 2.2

−1.809492	−1.037448	−0.392671	0.187575	0.9856874	1.4098688
−1.725369	−0.746903	−0.275223	0.4786776	0.9879926	1.4513166
−1.402125	−0.701965	−0.136124	0.7272926	0.9994073	1.594307
−1.137894	−0.556853	−0.095486	0.8280398	1.1616498	1.6920667
−1.038116	−0.424682	−0.017390	0.8382502	1.2449281	2.0837023

Describing the Distribution of Environmental Measurements

Probability distributions are mathematical functions that describe the probability that the value of x will lie in some interval for continuous distributions, or, x will equal some integer value for discrete distributions (e.g., integers only). There are two functional forms that are important in describing these distributions, the probability density function (PDF) and the cumulative distribution function (CDF). The PDF, which is written as f(X) can be thought of, in the case of continuous distributions, as providing information on the relative frequency or likelihood of different values of x, while for the case of discrete distributions it gives the probability, P, that x equals X; that is:

$$f(X) = P(x = X)$$ [2.14]

The CDF, usually written as F(X), always gives the probability that y is less than or equal to x; that is:

$$F(X) = P(x \leq X)$$ [2.15]

The two functions are related. For discrete distributions:

$$F(X) = \sum_{x=min}^{X} f(x)$$ [2.16]

For continuous distributions:

$$F(X) = \int_{x=min}^{X} f(x)\, dx$$ [2.17]

that is, the CDF is either the sum or integral of x between the minimum value for the distribution in question and the value of interest, X.

If one can find a functional form that they are willing to assume describes the underlying probability distribution for the observational set of measurements, then this functional form may be used as a model to assist with decision making based upon these measurements. The wise admonition of G. E. P. Box (1979) that "... all models are wrong but some are useful" should be kept firmly in mind when assuming the utility of any particular functional form. Techniques useful for judging the lack of utility of a functional form are discussed later in this chapter.

Some of the functional forms that traditionally have been found useful for continuous measurement data are the Gaussian or "normal" model, the "Student's t" distribution, and the log-normal model.

Another continuous model of great utility is the uniform distribution. The uniform model simply indicates that the occurrence of any measurement outcome within a range of possible outcomes is equally likely. Its utility derives from the fact that the CDF of any distribution is distributed as the uniform model. This fact will be exploited in discussing Bootstrap techniques in Chapter 6.

The Normal Distribution

The normal or Gaussian distribution is one of the historical cornerstones of statistical inference in that many broadly used techniques such as regression and analysis of variance (ANOVA) assume that the variation of measurement errors follows a normal distribution. The PDF for the normal distribution is given as:

$$f(x) = \frac{1}{\sigma(2\pi)^{1/2}} \exp[-1/2((x-\mu)/\sigma)^2] \qquad [2.18]$$

Here π is the numerical constant defined by the ratio of the circumference of circle to its diameter (≈ 3.14), exp is the exponential operator (exp $(Z) = e^Z$; e is the base of the natural logarithms (≈ 2.72)), and μ and σ are the parametric values for the mean and standard deviation, respectively. The CDF of the normal distribution does not have an explicit algebraic form and thus must be calculated numerically. A graph of the "standard" normal curve ($\mu = 0$ and $\sigma = 1$) is shown in Figure 2.3.

The standard form of the normal curve is important because if we subtract μ, the population mean, from each observation, and divide the result by σ, the sample standard deviation, the resulting transformed values have a mean of zero and a standard deviation of 1. If the parent distribution is normal the resulting standardized values should approximate a standard normal distribution. The standardization procedure is shown explicitly in Equation [2.19]. In this equation, Z is the standardized variate.

$$Z = (x - \mu)/\sigma \qquad [2.19]$$

The t Distribution

The t distribution, which is important in statistical estimation and hypothesis testing, is closely related to the normal distribution. If we have N observations from a normal distribution with parametric mean μ the t value is given by:

$$t = (\bar{x} - \mu)/S(\bar{x}) \qquad [2.20]$$

where

$$S(\bar{x}) = S/N^{1/2} \qquad [2.21]$$

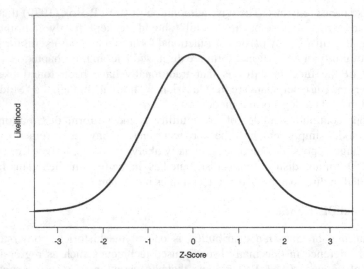

Figure 2.3   Graph of the PDF of a Standard Normal Curve
(Note that the likelihood is maximized at Z = 0, the
distribution mean.)

That is, $S(\bar{x})$ is the sample standard deviation divided by the square root of the sample size. A t distribution for a sample size of N is termed a t distribution on ν degrees of freedom, where $\nu = N - 1$ and is often written t_ν. Thus, for example, a t value based on 16 samples from a normal distribution would have a t_{15} distribution. The algebraic form of the t distribution is complex, but tables of the cumulative distribution function of t_ν are found in many statistics texts and are calculated by most statistical packages and some pocket calculators. Generally tabled values of t_ν are presented for $\nu = 1$ to $\nu = 30$ degrees of freedom and for probability values ranging from 0.90 to 0.9995. Many tables equivalently table 0.10 to 0.0005 for $1 - F(t_\nu)$. See Table 2.2 for some example t values. Note that Table 2.2 includes t_∞. This is the distribution of t for an infinite sample size, which is precisely equivalent to a normal distribution. As Table 2.2 suggests, for ν greater than 30 t tends toward a standard normal distribution.

The Log-Normal Distribution

Often chemical measurements exhibit a distribution with a long tail to the right. A frequently useful model for such data is the log-normal distribution. In such a distribution the logarithms of the x's follow a normal distribution. One can do logarithmic transformations in either log base 10 (often referred to as common logarithms, and written log(x)), or in log base e (often referred to as natural logarithms, and written as ln(x)). In our discussions we will always use natural logarithms because these are most commonly used in statistics. However, when confronted with "log-transformed data," the reader should always be careful to determine which logarithms are being used because log base 10 is also sometimes used. When dealing with log-normal statistical calculations all statistical tests are done with log-transformed observations, and assume a normal distribution.

Table 2.2
Some Values for t Distribution
(The entries in the body of the table are the t values.)

P Value	Degrees of Freedom (v)						
	1	2	5	10	20	30	∞
0.90	3.08	1.89	1.48	1.37	1.33	1.31	1.28
0.95	6.31	2.92	2.02	1.81	1.72	1.70	1.64
0.975	12.71	4.30	2.57	2.23	2.09	2.04	1.96
0.99	31.82	6.96	3.36	2.76	2.53	2.46	2.33
0.999	318.31	22.33	5.89	4.14	3.55	3.39	3.09
0.9995	636.62	31.6	6.87	4.59	3.85	3.65	3.29

A log-normal distribution, which corresponds to the exponential transformation of the standard normal distribution, is shown in Figure 2.4. An important feature of this distribution is that it has a long tail that points to the right and is thus termed "right skewed." The median and geometric mean for the example distribution are both 1.0, while the arithmetic mean is 1.65.

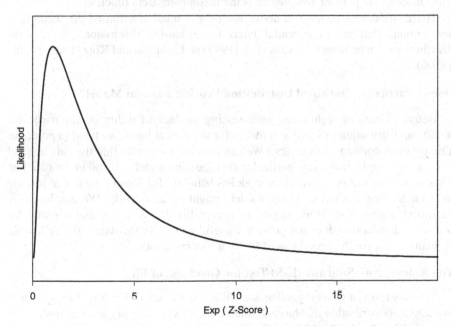

Figure 2.4   A Graph of the PDF of a Log-Normal Distribution Resulting from Exponentially Transforming the Z-Scores for a Standard Normal Curve

Some measurements such as counts of radioactive decay are usually expressed as events per unit time. The Poisson distribution is often useful in describing discrete measurements of this type. If we consider the number of measurements, x, out of a group of N measurements that have a particular property (e.g., they are above some "bright line" value such as effluent measurements exceeding a performance limitation), distributional models such as the binomial distribution models may prove useful. The functional forms of these are given below:

$$\text{Poisson Density, } f(x) = \frac{e^{-\lambda}\lambda^x}{x!} \qquad\qquad [2.22]$$

$$\text{Binomial Density, } f(x) = \binom{N}{x} p^x(1-p)^{N-x} \qquad [2.23]$$

In Equation 2.22, λ is the average number of events per unit time (e.g., counts per minute). In Equation 2.23, p is the probability that a single observation will be "positive" (e.g., exceed the "bright" line).

We may also be interested in the amount of time that will elapse until some event of interest will occur. These are termed "waiting time" distributions. When time is continuous, the exponential and Weibull distributions are well known. When time is discrete (e.g., number of measurement periods) waiting time is commonly described by the negative binomial distribution. An important aid in assigning a degree of confidence to the percent compliance is the Incomplete Beta function.

The distributions mentioned above are only a small fraction of the theoretical distributions that are of potential interest. Extensive discussion of statistical distributions can be found in Evans et al. (1993) and Johnson and Kotz (1969, 1970a, 1970b).

Does a Particular Statistical Distribution Provide a Useful Model?

Before discussing techniques for assessing the lack of utility of any particular statistical distribution to serve as a model for the data at hand, we need to point out a major short coming of statistics. We can never demonstrate that the data at hand arise as a sample from any particular distribution model. In other words, just because we can't reject a particular model as being useful doesn't mean that it is the only model that *is* useful. Other models might be as useful. We can however determine within a specified degree of acceptable decision risk that a particular statistical distribution does not provide a useful model for the data. The following procedures test for the "goodness of fit" of a particular model.

The Kolmogorov-Smirnov (K-S) Test for Goodness of Fit

The K-S test is a general goodness-of-fit test in the sense that it will apply to any hypothetical distribution that has a defined CDF, F(X). To apply this test in the case of a normal distribution:

A.   We sort our data from smallest to largest.

B. Next we calculate the standardized Z scores for each data value using Equation 2.19, with $\bar{x}$ and s substituted for μ and σ.

C. We then calculate the F(X) value for each Z-score either by using a table of the standard normal distribution or a statistics package or calculator that has built in normal CDF calculations. If we are using a table, it is likely that F(x) values are presented for $Z \geq 0$. That is, we will have Z values ranging from something like zero to 4, together with the cumulative probabilities (F(x)) associated with these Z values. For negative Z values, we use the relationship:

$$F(-Z) = 1 - F(Z)$$

that is, the P value associated with a negative Z value is equal to one minus the P value associated with the positive Z value of the same magnitude (e.g., $-1.5; 1.5$).

D. Next we calculate two measures of cumulative relative frequency:

$$C1 = RANK/N \quad \text{and} \quad C2 = (RANK - 1)/N$$

In both cases, N equals the sample size.

E. Now we calculate the absolute value of difference between C1 and F(Z) and C2 and F(Z) for each observation. That is:

$$DIFF1_i = |C1_i - F(Z)_i| \quad \text{and} \quad DIFF2_i = |C2_i - F(Z)_i|$$

F. Finally we select the largest of the $DIFF1_i$ and $DIFF2_i$ values. This is the value, D_{max}, used to test for significance (also called the "test statistic").

This calculation is illustrated in Table 2.3. Here our test statistic is 0.1124. This can be compared to either a standard probability table for the K-S statistic (Table 2.4) or in our example, Lilliefors modification of the K-S probability table (Lilliefors, 1967; Dallal and Wilkinson, 1986). The reason that our example uses Lilliefors modification of the K-S probabilities is that the K-S test compares a sample of measurements to a known CDF. In our example, F(X) was estimated using the sample mean x and standard deviation S. Lilliefors test corrects for the fact that F(X) is not really known *a priori*.

Dallal and Wilkinson (1986) give an analytic approximation to find probability values for Lilliefors test. For P< 0.10 and N between 5 and 100, this is given by:

$$P = \exp(-7.01256 \, D_{max}^2 (N + 2.78019) + 2.99587 \, D_{max}$$
$$(N + 2.78019)^{1/2} - 0.122119 + 0.974598/N^{1/2} \qquad [2.24]$$
$$+ 1.67997/N)$$

Table 2.3
A Sample Calculation for the Kolmogorov-Smirnov (K-S) Test for Goodness of Fit
(Maximum values for DIFF1 and DIFF2 are shown in ***bold italic*** type. The test statistic is 0.1124.)

Sample ID	Data Values	Rank	Rank/30	(Rank-1) /3030	Z-Score	Normal CDF: F(X)	DIFF1	DIFF2
1	0.88858	1	0.0333	0.0000	-2.2470	0.0123	0.0210	0.0123
1	1.69253	2	0.0667	0.0333	-1.5123	0.0652	0.0014	0.0319
1	1.86986	3	0.1000	0.0667	-1.3502	0.0885	0.0115	0.0218
1	1.99801	4	0.1333	0.1000	-1.2331	0.1088	0.0246	0.0088
1	2.09184	5	0.1667	0.1333	-1.1473	0.1256	0.0410	0.0077
1	2.20077	6	0.2000	0.1667	-1.0478	0.1474	0.0526	0.0193
1	2.25460	7	0.2333	0.2000	-0.9986	0.1590	0.0743	0.0410
1	2.35476	8	0.2667	0.2333	-0.9071	0.1822	***0.0845***	0.0511
2	2.55102	9	0.3000	0.2667	-0.7277	0.2334	0.0666	0.0333
1	2.82149	10	0.3333	0.3000	-0.4805	0.3154	0.0179	0.0154
2	3.02582	11	0.3667	0.3333	-0.2938	0.3845	0.0178	0.0511
2	3.05824	12	0.4000	0.3667	-0.2642	0.3958	0.0042	0.0292
1	3.12414	13	0.4333	0.4000	-0.2040	0.4192	0.0141	0.0192
1	3.30163	14	0.4667	0.4333	-0.0417	0.4834	0.0167	0.0500
1	3.34199	15	0.5000	0.4667	-0.0049	0.4981	0.0019	0.0314
1	3.53368	16	0.5333	0.5000	0.1703	0.5676	0.0343	0.0676
2	3.68704	17	0.5667	0.5333	0.3105	0.6219	0.0552	0.0886
1	3.85622	18	0.6000	0.5667	0.4651	0.6791	0.0791	***0.1124***
2	3.92088	19	0.6333	0.6000	0.5242	0.6999	0.0666	0.0999
2	3.95630	20	0.6667	0.6333	0.5565	0.7111	0.0444	0.0777
2	4.05102	21	0.7000	0.6667	0.6431	0.7399	0.0399	0.0733
1	4.09123	22	0.7333	0.7000	0.6799	0.7517	0.0184	0.0517
2	4.15112	23	0.7667	0.7333	0.7346	0.7687	0.0020	0.0354
2	4.33303	24	0.8000	0.7667	0.9008	0.8162	0.0162	0.0495
2	4.34548	25	0.8333	0.8000	0.9122	0.8192	0.0142	0.0192
2	4.35884	26	0.8667	0.8333	0.9244	0.8224	0.0443	0.0110
2	4.51400	27	0.9000	0.8667	1.0662	0.8568	0.0432	0.0098
2	4.67408	28	0.9333	0.9000	1.2125	0.8873	0.0460	0.0127
2	5.04013	29	0.9667	0.9333	1.5470	0.9391	0.0276	0.0057
2	5.33090	30	1.0000	0.9667	1.8128	0.9651	0.0349	0.0016

For sample sizes, K, greater than 100, Equation [2.24] is used with D_{max} replaced by D_{mod}:

$$D_{mod} = D_{max} \bullet (K/100)^{0.49} \qquad [2.25]$$

and N replaced by 100. For our example, Equation [2.24] gives P = 0.42, indication of a good fit to a normal distribution. Significant lack of fit is generally taken as P < 0.05.

Note that there are some instances where the K-S table would be appropriate. For example, if we had a large body of historical data on water quality that showed a log-normal distribution with logarithmic mean μ and logarithmic standard deviation σ and wanted to know if a set of current measurements followed the same distribution, we would use the K-S method with log-transformed sample data and with Z-scores calculated using μ and σ rather than x and S. More generally, if we wished to test a set of data against some defined cumulative distribution function, we would use the K-S table, not the Lilliefors approximation given in [2.24] and [2.25].

Normal Probability Plots

A second way to evaluate the goodness of fit to a normal distribution is to plot the data against the normal scores or Z scores expected on the basis of a normal distribution. Such plots are usually referred to as "normal probability plots," "expected normal scores plots," or "rankit plots." To make a normal probability plot:

1.   We sort the data from smallest to largest.

2.   We calculate the rank of the observation. Then, using Equation [2.6] we calculate the cumulative probability associated with each rank.

3.   We then calculate the F(X) value for each Z-score either by using a table of the standard normal distribution or a statistics package or calculator that has built in normal CDF calculations.

4.   We then plot the original data against the calculated Z-scores.

If the data are normal, the points in the plot will tend to fall along a straight line. Table 2.4 and Figure 2.5 show a normal probability plot using the same data as the K-S example. A goodness-of-fit test for a normal distribution can be obtained by calculating the correlation coefficient (see Chapter 4) and comparing it to the values given in Table 2.5 (Looney and Gulledge, 1985). In our example, the correlation coefficient (r) is 0.9896 (P $\approx$ 0.6), confirming the good fit of our example data to a normal distribution.

Table 2.4
A Sample Calculation for a Normal Probability Plot and Goodness-of-Fit Test

Sample ID	Data Values	Rank	(Rank-3/8)/ 30.25	Z-Score
1	0.88858	1	0.02066	−2.04028
1	1.69253	2	0.05372	−1.60982
1	1.86986	3	0.08678	−1.36087
1	1.99801	4	0.11983	−1.17581
1	2.09184	5	0.15289	−1.02411
1	2.20077	6	0.18595	−0.89292
1	2.25460	7	0.21901	−0.77555
1	2.35476	8	0.25207	−0.66800
2	2.55102	9	0.28512	−0.56769
1	2.82149	10	0.31818	−0.47279
2	3.02582	11	0.35124	−0.38198
2	3.05824	12	0.38430	−0.29421
1	3.12414	13	0.41736	−0.20866
1	3.30163	14	0.45041	−0.12462
1	3.34199	15	0.48347	−0.04144
1	3.53368	16	0.51653	0.04144
2	3.68704	17	0.54959	0.12462
1	3.85622	18	0.58264	0.20866
2	3.92088	19	0.61570	0.29421
2	3.95630	20	0.64876	0.38198
2	4.05102	21	0.68182	0.47279
1	4.09123	22	0.71488	0.56769
2	4.15112	23	0.74793	0.66800
2	4.33303	24	0.78099	0.77555
2	4.34548	25	0.81405	0.89292
2	4.35884	26	0.84711	1.02411
2	4.51400	27	0.88017	1.17581
2	4.67408	28	0.91322	1.36087
2	5.04013	29	0.94628	1.60982
2	5.33090	30	0.97934	2.04028

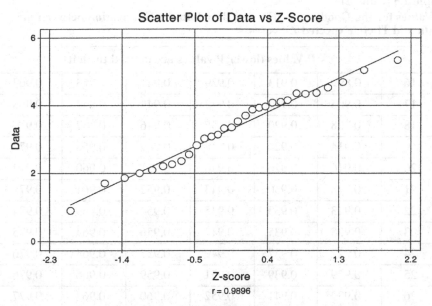

Figure 2.5 A Normal Scores Plot of the Data in Table 2.4

Table 2.5
P Values for the Goodness-of-Fit Test Based on the Correlation between the Data and Their Expected Z-Scores

	P Values (lower P values are toward the left)					
n	0.005	0.01	0.025	0.05	0.1	0.25
3	0.867	0.869	0.872	0.879	0.891	0.924
4	0.813	0.824	0.846	0.868	0.894	0.931
5	0.807	0.826	0.856	0.880	0.903	0.934
6	0.820	0.838	0.866	0.888	0.910	0.939
7	0.828	0.877	0.898	0.898	0.918	0.944
8	0.840	0.861	0.887	0.906	0.924	0.948
9	0.854	0.871	0.894	0.912	0.930	0.952
10	0.862	0.879	0.901	0.918	0.934	0.954
11	0.870	0.886	0.907	0.923	0.938	0.957
12	0.876	0.892	0.912	0.928	0.942	0.960
13	0.885	0.899	0.918	0.932	0.945	0.962
14	0.890	0.905	0.923	0.935	0.948	0.964
15	0.896	0.910	0.927	0.939	0.951	0.965

Table 2.5 (Cont'd)
P Values for the Goodness-of-Fit Test Based on the Correlation between the Data and Their Expected Z-Scores

	P Values (lower P values are toward the left)					
16	0.899	0.913	0.929	0.941	0.953	0.967
17	0.905	0.917	0.932	0.944	0.954	0.968
18	0.908	0.920	0.935	0.946	0.957	0.970
19	0.914	0.924	0.938	0.949	0.958	0.971
20	0.916	0.926	0.940	0.951	0.960	0.972
21	0.918	0.930	0.943	0.952	0.961	0.973
22	0.923	0.933	0.945	0.954	0.963	0.974
23	0.925	0.935	0.947	0.956	0.964	0.975
24	0.927	0.937	0.949	0.957	0.965	0.976
25	0.929	0.939	0.951	0.959	0.966	0.976
26	0.932	0.941	0.952	0.960	0.967	0.977
27	0.934	0.943	0.953	0.961	0.968	0.978
28	0.936	0.944	0.955	0.962	0.969	0.978
29	0.939	0.946	0.956	0.963	0.970	0.979
30	0.939	0.947	0.957	0.964	0.971	0.979
31	0.942	0.950	0.958	0.965	0.972	0.980
32	0.943	0.950	0.959	0.966	0.972	0.980
33	0.944	0.951	0.961	0.967	0.973	0.981
34	0.946	0.953	0.962	0.968	0.974	0.981
35	0.947	0.954	0.962	0.969	0.974	0.982
36	0.948	0.955	0.963	0.969	0.975	0.982
37	0.950	0.956	0.964	0.970	0.976	0.983
38	0.951	0.957	0.965	0.971	0.976	0.983
39	0.951	0.958	0.966	0.971	0.977	0.983
40	0.953	0.959	0.966	0.972	0.977	0.984
41	0.953	0.960	0.967	0.973	0.977	0.984
42	0.954	0.961	0.968	0.973	0.978	0.984
43	0.956	0.961	0.968	0.974	0.978	0.984
44	0.957	0.962	0.969	0.974	0.979	0.985
45	0.957	0.963	0.969	0.974	0.979	0.985

Table 2.5 (Cont'd)
P Values for the Goodness-of-Fit Test Based on the Correlation between the Data and Their Expected Z-Scores

	P Values (lower P values are toward the left)					
46	0.958	0.963	0.970	0.975	0.980	0.985
47	0.959	0.965	0.971	0.976	0.980	0.986
48	0.959	0.965	0.971	0.976	0.980	0.986
49	0.961	0.966	0.972	0.976	0.981	0.986
50	0.961	0.966	0.972	0.977	0.981	0.986
55	0.965	0.969	0.974	0.979	0.982	0.987
60	0.967	0.971	0.976	0.980	0.984	0.988
65	0.969	0.973	0.978	0.981	0.985	0.989
70	0.971	0.975	0.979	0.983	0.986	0.990
75	0.973	0.976	0.981	0.984	0.987	0.990
80	0.975	0.978	0.982	0.985	0.987	0.991
85	0.976	0.979	0.983	0.985	0.988	0.991
90	0.977	0.980	0.984	0.986	0.988	0.992
95	0.979	0.981	0.984	0.987	0.989	0.992
100	0.979	0.982	0.985	0.987	0.989	0.992

Testing Goodness of Fit for a Discrete Distribution: A Poisson Example

Sometimes the distribution of interest is discrete. That is, the object of interest is counts, not continuous measurements. An actual area where such statistics can be of importance is in studies of bacterial contamination. Let us assume that we have a set of water samples, and have counted the number of bacteria in each sample. For such a problem, a common assumption is that the distribution of counts across samples follows a Poisson distribution given by Equation [2.22].

If we simply use $\bar{x}$ calculated from our samples in place of λ in [2.22], and calculate $f(x)$ for each x, these $f(x)$ can then be used in a chi-squared goodness-of-fit test to assess whether or not the data came from a Poisson distribution.

Table 2.6 shows a goodness-of-fit calculation for some hypothetical bacterial count data. Here we have a total of 100 samples with bacterial counts ranging from 7 to 25 (Column 1). Column 2 gives the numbers of samples that had different counts. In Column 3 we show the actual frequency categories to be used in our goodness-of-fit test. Generally for testing goodness of fit for discrete distributions, we define our categories so that the *expected* number (not the observed number) of observations under our null hypothesis (H_0: "the data are consistent with a Poisson distribution") is at least 5. Since we have 100 total observations, we select categories so that the probability for the category is at least 0.05. Column 4 shows the category observed frequencies, Column 5 shows our category probabilities. Note that for a category with multiple x values (e.g., <10) the probability is given as the sum of the probabilities of the individual x's. For the 10 or less category, this is the sum of the f(x) values for x = 0 to x = 10, that is, the CDF, F (10). For the 20+ category the P value is most easily obtained as:

$$P = 1 - F(19) \qquad\qquad [2.26]$$

When calculating probabilities for probabilities for generation expected frequencies, we must consider all possible outcomes, not just those observed. Thus, we observed no samples with counts less than 7, but the probabilities for counts 0–6 are incorporated in the probability of the less than 10 category. Similarly, we observed no samples with counts greater than 25 (or equal to 20 or 21 for that matter). The probability for the 20+ category is 0.1242.

Column 6 shows our expected frequencies, which are given by the numbers in Column 5 times our sample size N, in our sample N = 100. In Column 7 we show our calculated chi-squared statistics. The chi-squared statistic, often written as "χ^2" is given by:

$$\chi^2 = (O - E)^2 / E \qquad\qquad [2.27]$$

Here O is the observed frequency (Column 4) and E is the expected frequency (Column 6). The actual test statistic is the sum of the chi-squared values for each category. This is compared to the chi-squared values found in Table 2.7. The degrees of freedom (ν; see our discussion of the t-distribution) are given by M – 2, where M is the number of categories involved in calculation of the overall chi-squared statistic. Here M = 11. Consulting Table 2.7 we see that a chi-squared value of 10.1092 (χ^2 = 10.1092) with 9 degrees of freedom ($\nu = 9$) has a tail probability of P > 0.10. Since this probability is greater than an acceptable decision error of 0.05, we conclude that the Poisson distribution reasonably describes the distribution of our data.

Table 2.6
Testing Goodness of Fit to a Poisson Distribution

Number of Bacteria	Samples with Count	Categories Used for Test	Observed Frequency	Category Probabilities	Expected Frequency	Chi-Squared Statistics
7	2					
8	3					
9	2					
10	3	<10	10	0.1190	11.8950	0.3019
11	7	7	7	0.0665	6.6464	0.0188
12	7	7	7	0.0830	8.3025	0.2043

Table 2.6 (Cont'd)
Testing Goodness of Fit to a Poisson Distribution

Number of Bacteria	Samples with Count	Categories Used for Test	Observed Frequency	Category Probabilities	Expected Frequency	Chi-Squared Statistics
13	7	7	7	0.0957	9.5734	0.6918
14	12	12	12	0.1025	10.2504	0.2986
15	14	14	14	0.1024	10.2436	1.3775
16	10	10	10	0.0960	9.5969	0.0169
17	8	8	8	0.0846	8.4622	0.0252
18	12	12	12	0.0705	7.0472	3.4809
19	7	7	7	0.0556	5.5598	0.3730
22	3	20+	6	0.1242	12.4220	3.3201
23	1					
24	1					
25	1					
Totals	100		100	1.0	100	10.1092

Table 2.7
Critical Values of the χ^2 Distribution

df	Tail Probabilities								
	0.250	0.200	0.150	0.100	0.050	0.025	0.010	0.005	0.001
1	1.323	1.642	2.072	2.706	3.841	5.024	6.635	7.879	10.828
2	2.773	3.219	3.794	4.605	5.991	7.378	9.210	10.597	13.816
3	4.108	4.642	5.317	6.251	7.815	9.348	11.345	12.838	16.266
4	5.385	5.989	6.745	7.779	9.488	11.143	13.277	14.860	18.467
5	6.626	7.289	8.115	9.236	11.070	12.833	15.086	16.750	20.515
6	7.841	8.558	9.446	10.645	12.592	14.449	16.812	18.548	22.458
7	9.037	9.803	10.748	12.017	14.067	16.013	18.475	20.278	24.322
8	10.219	11.030	12.027	13.362	15.507	17.535	20.090	21.955	26.124
9	11.389	12.242	13.288	14.684	16.919	19.023	21.666	23.589	27.877
10	12.549	13.442	14.534	15.987	18.307	20.483	23.209	25.188	29.588
11	13.701	14.631	15.767	17.275	19.675	21.920	24.725	26.757	31.264
12	14.845	15.812	16.989	18.549	21.026	23.337	26.217	28.300	32.909
13	15.984	16.985	18.202	19.812	22.362	24.736	27.688	29.819	34.528
14	17.117	18.151	19.406	21.064	23.685	26.119	29.141	31.319	36.123

Table 2.7 (Cont'd)
Critical Values of the χ^2 Distribution

df	Tail Probabilities								
	0.250	0.200	0.150	0.100	0.050	0.025	0.010	0.005	0.001
15	18.245	19.311	20.603	22.307	24.996	27.488	30.578	32.801	37.697
16	19.369	20.465	21.793	23.542	26.296	28.845	32.000	34.267	39.252
17	20.489	21.615	22.977	24.769	27.587	30.191	33.409	35.718	40.790
18	21.605	22.760	24.155	25.989	28.869	31.526	34.805	37.156	42.312
19	22.718	23.900	25.329	27.204	30.144	32.852	36.191	38.582	43.820
20	23.828	25.038	26.498	28.412	31.410	34.170	37.566	39.997	45.315
25	29.339	30.675	32.282	34.382	37.652	40.646	44.314	46.928	52.620
30	34.800	36.250	37.990	40.256	43.773	46.979	50.892	53.672	59.703
40	45.616	47.269	49.244	51.805	55.758	59.342	63.691	66.766	73.402
50	56.334	58.164	60.346	63.167	67.505	71.420	76.154	79.490	86.661
60	66.981	68.972	71.341	74.397	79.082	83.298	88.379	91.952	99.607
70	77.577	79.715	82.255	85.527	90.531	95.023	100.425	104.215	112.317
80	88.130	90.405	93.106	96.578	101.879	106.629	112.329	116.321	124.839
90	98.650	101.054	103.904	107.565	113.145	118.136	124.116	128.299	137.208
100	109.141	111.667	114.659	118.498	124.342	129.561	135.807	140.169	149.449

Constructed using SAS (1990) function for generating quantiles of the chi-square distribution, CINV.

Confidence Intervals

A confidence interval is defined as an interval that contains the true value of the statistic of interest with some probability (Hahn and Meeker, 1991). Thus a 95% confidence interval has a 95% probability of containing the true population mean. In our preceding discussion we considered a number of measures for location (mean, geometric mean, median) and dispersion (arithmetic and geometric standard deviations and variances; interquartile range). In fact confidence intervals can be calculated for all of these quantities (Hahn and Meeker, 1991), but we will focus on confidence intervals for the arithmetic and geometric mean.

Confidence Intervals from the Normal Distribution

In our earlier discussion of distributions we said that for a normal distribution one can use either the standard normal distribution (Z-scores) or the t distribution (for smaller samples) to calculate the probability of different values of x. Recall that Equation [2.20] gave the formula for the t statistic as $t = (\bar{x} - \mu)/S(\bar{x})$. ($S(\bar{x})$ is defined by Equation 2.21.) We do not know μ, but can use t to get a confidence interval for μ.

An upper confidence bound, $U(\bar{x})$ for a two-sided probability interval of width $(1 - \alpha)$ is given by:

$$U(\bar{x}) = \bar{x} + t_{v,\,(1-\alpha/2)}\, S(\bar{x}) \qquad\qquad [2.28]$$

that is, the upper bound is given by the sample mean *plus* the t statistic with v degrees of freedom, for probability $(1 - \alpha/2)$ times the standard error of the sample mean.

Similarly a lower confidence bound, $L(\bar{x})$ for a two-sided probability interval of width $(1 - \alpha)$ is given by:

$$L(\bar{x}) = \bar{x} - t_{v,\,(1-\alpha/2)}\, S(\bar{x}) \qquad\qquad [2.29]$$

that is, the lower bound is given by the sample mean *minus* the t statistic with v degrees of freedom, for probability $(1 - \alpha/2)$ times the standard error of the sample mean.

The values $L(\bar{x})$ and $U(\bar{x})$ are the ends of a two-sided interval that contains the true population mean, μ, with probability $(1 - \alpha)$.

We note that there are also one-sided confidence intervals. For the t interval discussed above, a one-sided $(1 - \alpha)$ upper interval is given by:

$$U(\bar{x}) = \bar{x} + t_{v,\,(1-\alpha)}\, S(\bar{x}) \qquad\qquad [2.30]$$

The only difference is that instead of $t_{v,\,(1-\alpha/2)}$, we use $t_{v,\,(1-\alpha)}$. In English, the interpretation of [2.30] is that the true population mean, μ, is less than $U(\bar{x})$ with probability $(1 - \alpha)$. One-sided upper bounds are common in environmental sampling because we want to know how high the concentration might be, but are not interested in how low it might be. It is important to know if confidence intervals are one- or two-sided. Let us consider Table 2.2; if we had a sample of size 11 ($v = 10$), and were interested in a 0.95 one-sided interval, we would use $t = 1.81$ ($t_{10, 0.95}$), but if we were calculating the upper bound on a two-sided interval, we would use $t = 2.23$ ($t_{10, 0.975}$). Some statistical tables assume that one is calculating a two-sided interval, and thus give critical values of $(1 - \alpha/2)$ for $P = \alpha$.

As noted above upper bounds on the sample mean are the usual concern of environmental quality investigations. However, the distribution of chemical concentrations in the environment is often highly skewed. In this case one sometimes assumes that the data are log-normally distributed and calculates an interval based on log-transformed data. That is, one calculates $\bar{x}$ and $S(\bar{x})$ for the logarithms of the data, and then obtains the $U(\bar{x})$ and, if desired $L(\bar{x})$, using the t or normal distribution. Confidence bounds in original units can then be obtained as:

$$U_o = \exp(U(\bar{x})) \quad \text{and} \quad L_o = \exp(L(\bar{x})) \qquad\qquad [2.31]$$

The problem here is that $\exp(\bar{x})$ is the geometric mean. Thus U_o and L_o are confidence bounds on the geometric mean. As we saw above, the geometric mean is always less than the arithmetic mean. Moreover, for strongly right-skewed data (e.g., Figure 2.4) the difference can be considerable.

Mean and Variance Relationships for Log-Normal Data

Sometimes when one is dealing with data collected by others, one has only summary statistics, usually consisting of means and standard deviations. If we believe that the data are in fact log-normal, we can convert between arithmetic and geometric means and variances (and hence standard deviations because the latter is the square root of the former). For our discussion we will take $\bar{x}$ and S as the mean and standard deviation of log-transformed sample observations; M and SD as the arithmetic mean and standard deviation of the sample; and GM and GSD as the geometric mean and standard deviation of the sample.

If we are given GM and GSD, we can find $\bar{x}$ and S as:

$$\bar{x} = \ln(GM) \text{ and } S = \ln(GSD) \tag{2.32}$$

(Remember that the relationships in [2.32] hold only if we are using natural logarithms.)

If we know $\bar{x}$ and S, we can find M by the approximation:

$$M \approx \exp(\bar{x} + (S^2/2)) \tag{2.33}$$

We can also find SD as

$$SD \approx M \, (\exp(S^2) - 1)^{1/2} \tag{2.34}$$

If instead we begin with M and SD, we can find S as:

$$S^2 \approx \ln((SD^2/M^2) + 1) \tag{2.35}$$

We can then find $\bar{x}$ as:

$$\bar{x} \approx \ln(M) - (S^2/2) \tag{2.36}$$

Once we have $\bar{x}$ and S^2, we can find GM and GSD using the relationships shown in [2.32]. The relationships shown in [2.32] through [2.36] allow one to move between geometric and arithmetic statistics, provided the data are really log-normal. This can be useful in the sense that we can calculate approximate confidence intervals for geometric means when only arithmetic statistics are reported. Similarly, if one has only geometric statistics, one might wish to get an estimate of the arithmetic mean. A more extensive discussion of mean and standard deviation estimators for log-normal data can be found in Gilbert (1987).

Other Intervals for Sample Means

For many environmental quality problems the mean of interest is the arithmetic mean because long-term human exposure in a given environment will be well approximated by the arithmetic mean. Thus the question arises, "If the data are highly right skewed, how does one calculate an upper bound on the arithmetic mean?" The EPA recommends use of the 95% upper bound on the mean of the data as the "exposure point concentration" (EPC) in human health risk assessments (USEPA, 1989, 1992). Further, EPA recommends that, in calculating that 95% UCL for data that are log-normally distributed, the H-statistic procedure developed by Land (1975) be used (USEPA, 1992). Unfortunately, if the sample data are overdispersed (spread out) relative to a log-normal distribution, the Land method will give upper bounds that are much too large relative to the correct value (Singh et al., 1997, 1999; Ginevan and Splitstone, 2002). Moreover, almost all environmental data are likely to be spread out relative to a log-normal distribution, and yet such a distribution may appear log-normal on statistical tests (Ginevan and Splitstone, 2002). In such cases, the Land procedure can produce an EPC that is far higher than the true upper bound. For example, Ginevan and Splitstone (2002) showed that the Land upper bound can overestimate the true upper bound by a factor of 800, and that overestimates of the order of 30- to 50-fold were common.

The potential bias in the Land procedure is so severe that all authors who have carefully studied its behavior have recommended that it should be used cautiously with environmental contamination data (Singh et al., 1997, 1999; Ginevan and Splitstone, 2001; Gilbert, 1993). Earlier recommendations have said that the procedure may be used if the data are fit well by a log-normal distribution (Singh et al., 1997; Ginevan, 2001). However because our more recent work (Ginevan and Splitstone, 2002) suggests that lack of fit to a log-normal distribution may be nearly impossible to detect, we take the stronger position that the Land procedure should never be used with environmental contamination data. Thus, we do not discuss its calculation here.

Confidence intervals based on Chebyshev's inequality (Singh et al., 1997) have also been suggested for environmental contamination data. This inequality states that for any random variable X that has a finite variance, σ^2, it is true that:

$$P(|X - \mu| \geq k\sigma) \leq 1/k^2 \qquad [2.37]$$

(Hogg and Craig, 1995). This is true regardless of the distribution involved. Thus the suggestion that this inequality can be used to construct robust confidence intervals. A couple of problems make this idea questionable. First, [2.37] depends on knowing the parametric mean and variance (μ and σ^2), so if we substitute the sample mean and variance ($\bar{x}$ and S^2) [2.31] is no longer strictly true. Moreover, intervals based on Chebyshev's Inequality can be extremely broad. Consider [2.37] for k = 1. In English, [2.37] would say that the sample mean is within 1 standard deviation of the parametric mean with a probability of 1 or less. This is true but not helpful (all events have a probability of one or less). In their discussion of Chebyshev's Inequality, Hogg and Craig (1995) state: "[Chebyshev bounds] are not necessarily close to the exact probabilities and accordingly we ordinarily do not use the [Chebyshev] theorem to approximate a probability. The principal uses of the theorem and a special case of it are in theoretical discussions ..."; that is, its usefulness is in proving theorems, not in constructing confidence intervals.

One method of constructing nonparametric confidence intervals for the mean is the procedure known as bootstrapping or resampling (Efron and Tibshirani, 1993). In our view, this is the best approach to constructing confidence intervals for means when the data are not normally distributed as well as for performing nonparametric hypothesis tests. Because of its importance we devote all of Chapter 6 to this procedure.

Useful Bounds for Population Percentiles

One question that can be answered in a way that does not depend on a particular statistical distribution is "How much of the population of interest does my sample cover?" That is, if we have a sample of 30, we expect to have a 50-50 chance of having the largest sample value represent the 97.7th percentile of the population sampled and a 95% chance of having the largest sample value represent the 90.5th percentile of the population sampled. These numbers arise as follows:

1. If we have N observations and all are less than some cumulative probability $P = F(x)$, it follows that the probability of this event, Q, can be calculated as:

$$Q = P^N \qquad [2.38]$$

2. Now we can rearrange our question to ask "if $Q = Z$, what is the P value associated with this Q?" In other words, if

$$P = Q^{1/N} \qquad [2.39]$$

then we have a 50-50 chance of seeing the 97.7th percentile. Similarly, for $Q = 0.05$, we have a 95% chance of seeing the 90.5th percentile.

More generally, if we are interested in a $1 - \alpha$ lower bound on P we take Q equal to α. That is, if we wanted a 90% lower bound for P we would take Q equal to 0.10. Note that we are not usually interested in upper bounds because we want to know what percentile of the population we are likely to observe. However, if we did want to calculate a $1 - \alpha$ upper bound we would take Q equal to $1 - \alpha$. That is if we wanted a 90% upper bound, we would take Q equal to 0.90. These kinds of calculations, which are based on an area of statistics known as "order statistics" (David, 1981), can be useful in assessing how extreme the largest value in the sample actually is and can thus provide an assessment of the likelihood that our sample includes really extreme values.

References

Box, G. E. P., 1979, "Robustness in the Strategy of Scientific Model Building," *Robustness in Statistics*, eds. R. L. Launer and G. N. Wilkinson, Academic Press, pp. 201–236.

D'Agostino, R. B. and Stephens, M. A., (eds.), 1986, *Goodness-of-Fit Techniques*, Marcel Dekker, New York.

Dallal, G. E. and Wilkinson, L., 1986, "An Analytic Approximation to the Distribution of Lilliefor's Test Statistic for Normality," *American Statistician*. 40: 294–296.

David, H. A, 1981, *Order Statistics*, John Wiley, New York.

Efron, B. and Tibshirani, R. J., 1993, *An Introduction to the Bootstrap*, Chapman and Hall, London.

Evans, M., Hastings, N., and Peacock, B., 1993, *Statistical Distributions*, John Wiley and Sons, New York.

Gilbert, R. O., 1987, Statistical Methods for Environmental Pollution Monitoring. Van Nostrand Reinhold, New York.

Ginevan, M. E. and Splitstone, D. E., 2002, "Bootstrap Upper Bounds for the Arithmetic Mean of Right-Skewed Data, and the Use of Censored Data," *Environmetrics* (in press).

Ginevan, M. E., 2001, "Using Statistics in Health and Environmental Risk Assessments," *A Practical Guide to Understanding, Managing, and Reviewing Environmental Risk Assessment Reports*, eds. S. L. Benjamin and D. A. Belluck, Lewis Publishers, New York, pp. 389–411.

Hahn, G. J. and Meeker, W. Q., 1991, *Statistical Intervals: A Guide for Practitioners*, New York, John Wiley.

Hogg, R. V. and Craig, A. T., 1995, *An Introduction to Mathematical Statistics, 5th Edition*, Prentice Hall, Englewood Cliffs, NJ.

Johnson, N. L. and Kotz, S., 1969, *Distributions in Statistics: Discrete Distributions*, John Wiley and Sons, New York.

Johnson, N. L. and Kotz, S., 1970a, *Distributions in Statistics: Continuous Univariate Distributions — 1*, John Wiley and Sons, New York.

Johnson, N. L. and Kotz, S., 1970b, *Distributions in Statistics: Continuous Univariate Distributions — 2*, John Wiley and Sons, New York.

Land, C. E, 1975, "Tables of Confidence Limits for Linear Functions of the Normal Mean and Variance," *Selected Tables in Mathematical Statistics*, Vol. III, American Mathematical Society, Providence, RI, pp. 385–419.

Lillicfors, H. W., 1967, "On the Kolmogorov-Smirnov Tests for Normality with Mean and Variance Unknown," *Journal of the American Statistical Association*, 62: 399–402.

Looney, S. W., and Gulledge, Jr., T. R., 1985, "Use of the Correlation Coefficient with Normal Probability Plots," *American Statistician*, 39: 75–79.

Moore, D. S. and McCabe, G. P., 1993, *Introduction to the Practice of Statistics, 2nd ed.*, Freeman, New York, p. 1.

SAS Institute Inc., 1990, *SAS Language Reference, Version 6, First Edition*, SAS Institute, Cary, NC.

Sheskin, D. J., 2000, *Handbook of Parametric and Nonparametric Statistical Procedures.* Second Edition. CRC Press, Boca Raton, FL.

Singh, A. K., Singh, A., and Engelhardt, M., 1997, *The Lognormal Distribution in Environmental Applications.* EPA/600/R-97/006, December, 19p.

USEPA, 1989, *Risk Assessment Guidance for Superfund: Human Health Evaluation Manual – Part A*, Interim Final, United States Environmental Protection Agency, Office of Emergency and Remedial Response. Washington, D.C.

USEPA, 1992, *Supplemental Guidance to RAGS; Calculating the Concentration Term, Volume 1, Number 1*, Office of Emergency and Remedial Response, Washington, D.C., NTIS PE92-963373.

CHAPTER 3

Hypothesis Testing

"Was it due to chance, or something else? Statisticians have
invented *tests of significance* to deal with this sort of question."
(Freedman, Pisani, and Purves, 1997)

Step 5 of EPA's DQO process translates the broad questions identified in Step 2
into specific testable statistical hypothesis. Examples of the broad questions might
be the following.

- Does contamination at this site pose a risk to health and the environment?
- Is the permitted discharge in compliance with applicable limitations?
- Is the contaminant concentration significantly above background levels?
- Have the remedial cleanup goals been achieved?

The corresponding statements that may be subject to statistical evaluation might be
the following:

- The median concentration of acrylonitrile in the upper foot of soil at this
  residential exposure unit is less than or equal to 5 mg/kg?

- The 30-day average effluent concentration of zinc if the wastewater discharge
  from outfall 012 is less than or equal to 137 µg/l?

- The geometric mean concentration of lead in the exposure unit is less than or
  equal to that found in site specific background soil?

- The concentration of thorium in surface soil averaged over a 100-square-
  meter remedial unit is less than or equal to 10 picocuries per gram?

These specific statements, which may be evaluated with a statistical test of
significance, are called the *null hypothesis* often symbolized by H_0. It should be
noted that all statistical tests of significance are designed to assess the strength of
evidence *against* the null hypothesis.

Francis Y. Edgeworth (1845–1926) first clearly exposed the notion of
significance tests by considering, "Under what circumstances does a difference in
[calculated] figures correspond to a difference of fact" (Moore and McCabe, 1993,
p. 449, Stigler, 1986, p. 308). In other words, under what circumstances is an
observed outcome significant. These circumstances occur when *the outcome
calculated from the available evidence (the observed data) is not likely to have
resulted if the null hypothesis were correct.* The definition of what is *not likely* is
entirely up to us, and can always be fixed for any statistical test of significance. It is
very analogous to the beyond-a-reasonable-doubt criteria of law where we get to
quantify ahead of time the maximum probability of the outcome that represents a
reasonable doubt.

49

Step 6 of the DQO process refers to the specified maximum reasonable doubt probability as the probability of *false positive decision error*. Statisticians simply refer to this decision error of rejecting the null hypothesis, H_0, when it is in fact true as an error of *Type I*. The specified probability of committing a Type I error is usually designated by the Greek letter α.

The specification of α depends largely on the consequences of deciding the null hypothesis is false when it is in fact true. For instance, if we conclude that the median concentration of acrylonitrile in the soil of the residential exposure unit exceeds 5 mg/kg when it is in truth less than 5 mg/kg, we would incur the cost of soil removal and treatment or disposal. These costs represent real out-of-pocket dollars and would likely have an effect that would be noted on a firm's SEC Form 10Q. Therefore, the value assigned to α should be small. Typically, this represents a one-in-twenty chance ($\alpha = 0.05$) or less.

Every thesis deserves an antithesis and null hypotheses are no different. The *alternate hypothesis*, $\mathbf{H_1}$, is a statement that we assume to be true in lieu of H_0 when it appears, based upon the evidence, that H_0 is not likely. Below are some alternate hypotheses corresponding to the H_0's above.

- The median concentration of acrylonitrile in the upper foot of soil at this residential exposure unit is greater than 5 mg/kg.

- The 30-day average effluent concentration of zinc if the wastewater discharge from outfall 012 exceeds 137 µg/l.

- The geometric mean concentration of lead in the exposure unit is greater than the geometric mean concentration found in site specific background soil.

- The concentration of thorium in surface soil averaged over a 100-square-meter remedial unit is greater than 10 picocuries per gram.

We have controlled and fixed the error associated with choosing the alternate hypothesis, H_1, when the null hypothesis, H_0, is indeed correct. However, we must also admit that the available evidence may favor the choice of H_0 when, in fact, H_1 is true. DQO Step 6 refers to this as a *false negative decision error*. Statisticians call this an error of Type II and the magnitude of the Type II error is usually symbolized by Greek letter β. β is a function of both the sample size and the degree of true deviation from the conditions specified by H_0, given that α is fixed.

There are consequences associated with committing a Type II error that ought to be considered, as well as those associated with an error of Type I. Suppose that we conclude that concentration of thorium in surface soil averaged over a 100-square-meter remedial unit is less than 10 picocuries per gram; that is, we adopt H_0. Later, during confirmatory sampling it is found that the average concentration of thorium is greater than 10 picocuries per gram. Now the responsible party may face incurring costs for a second mobilization; additional soil excavation and disposal; and, a second confirmatory sampling. β specifies the probability of incurring these costs.

The relative relationship of Type I and Type II errors and the null hypothesis is summarized in Table 3.1.

Table 3.1
Type I and II Errors

Unknown Truth	Decision Made	
	Accept H_0	Reject H_0
H_0 True	No Error	Type I Error (α)
H_0 False	Type II Error (β)	No Error

Rarely, in the authors' experience, do parties to environmental decision making pay much, if any, attention to the important step of specifying the tolerable magnitude of decision errors. The magnitude of both the Type I and Type II error, α and β, has a direct link to the determination of the number of the samples to be collected. Lack of attention to this important step predictably results in multiple cost overruns.

Following are several examples that illustrate the concepts involved with the determination of statistical significance in environmental decision making via hypothesis evaluation. These examples provide illustration of the concepts discussed in this introduction.

Tests Involving a Single Sample

The simplest type of hypothesis test is one where we wish to compare a characteristic of a population against a fixed standard. Most often this characteristic describes the "center" of the distribution of concentration, the mean or median, over some physical area or span of time. In such situations we estimate the desired characteristic from one or more representative statistical samples of the population. For example, we might ask the question "Is the median concentration of acrylonitrile in the upper foot of soil at this residential exposure unit less than or equal to 5 mg/kg."

Ignoring for the moment the advice of the DQO process, the management decision was to collect 24 soil samples. The results of this sampling effort appear in Table 3.2.

Using some of the techniques described in the previous chapter, it is apparent that the distribution of the concentration data, y, is skewed. In addition it is noted that the log-normal model provides a reasonable model for the data distribution. This is fortuitous, for we recall from the discussion of confidence intervals that for a log-normal distribution, half of the samples collected would be expected to have concentrations above, and half below, the geometric mean. Therefore, in expectation the geometric mean and median are the same. This permits us to formulate hypotheses in terms of the logarithm of concentration, x, and apply standard statistical tests of significance that appeal to the normal theory of errors.

Table 3.2
Acrylonitrile in Samples from Residential Exposure Unit

Sample Number	Acrylonitrile (mg/kg, y)	x = ln(y)	Above 5 mg/kg
S001	45.5	3.8177	Yes
S002	36.9	3.6082	Yes
S003	25.6	3.2426	Yes
S004	36.5	3.5973	Yes
S005	4.7	1.5476	No
S006	14.4	2.6672	Yes
S007	8.1	2.0919	Yes
S008	15.8	2.7600	Yes
S009	9.6	2.2618	Yes
S010	12.4	2.5177	Yes
S011	3.7	1.3083	No
S012	2.6	0.9555	No
S013	8.9	2.1861	Yes
S014	17.6	2.8679	Yes
S015	4.1	1.4110	No
S016	5.7	1.7405	Yes
S017	44.2	3.7887	Yes
S018	16.5	2.8034	Yes
S019	9.1	2.2083	Yes
S020	23.5	3.1570	Yes
S021	23.9	3.1739	Yes
S022	284	5.6507	Yes
S023	7.3	1.9879	Yes
S024	6.3	1.8406	Yes
Mean, $\bar{x}$ =		2.6330	
Std. deviation, S =		1.0357	
Number greater than 5 mg/kg, **w** = 20			

Consider a null, H_0, and alternate, H_1, hypothesis pair stated as:

H_0: Median acrylonitrile concentration is less than or equal to 5 mg/kg;

H_1: Median acrylonitrile concentration is greater than 5 mg/kg;

Given the assumption of the log-normal distribution these translate into:

H_0: The mean of the log acrylonitrile concentration, μ_x, is less than or equal to ln(5 mg/kg);

H_1: The mean of the log acrylonitrile concentration, μ_x, is greater than ln(5 mg/kg).

Usually, these statements are economically symbolized by the following shorthand:

$$H_0: \mu_x \le \mu_0 \,(= \ln(5 \text{ mg/kg}) = 1.6094);$$

$$H_1: \mu_x > \mu_0 \,(= \ln(5 \text{ mg/kg}) = 1.6094).$$

The sample mean $\bar{x}$ standard deviation (S), sample size (N), and population mean μ, hypothesized in H_0 are connected by the student's "t" statistics introduced in Equation [2.20]. Assuming that we are willing to run a 5% chance ($\alpha = 0.05$) of rejecting H_0 when it is true, we may formulate a decision rule. That rule is "**we will reject H_0 if the calculated value of t is greater than the 95th percentile of the t distribution with 23 degrees of freedom.**" This value, $t_{v=23,\,0.95} = 1.714$, may be found by interpolation in Table 2.2 or from the widely published tabulation of the percentiles of Student's t-distribution such as found in *Handbook of Tables for Probability and Statistics* from CRC Press:

$$t - \frac{\bar{x} - \mu_0}{S/\sqrt{N}} = \frac{2.6330 - 1.6094}{1.0357/\sqrt{24}} = 4.84 \qquad [3.1]$$

Clearly, this value is greater than $t_{v=23,\,0.95} = 1.714$ and we reject the hypothesis that the median concentration in the exposure area is less than or equal to 5 mg/kg.

Alternately, we can perform this test by simply calculating a 95% one-sided lower bound on the geometric mean. If the target concentration of 5 mg/kg lies above this limit, then we cannot reject H_0. If the target concentration of 5 mg/kg lies below this limit, then we must reject H_0.

This confidence limit is calculated using the relationship given by Equation [2.29] modified to place all of the Type I error in a single tail of the "t" distribution to accommodate the single-sided nature of the test. The test is single sided simply because if the true median is below 5 mg/kg, we don't really care how much below.

$$L(\bar{x}) = \bar{x} - t_{v,(1-\alpha)}\, S/\sqrt{N}$$

$$L(\bar{x}) = 2.6330 - 1.714 \bullet 1.0357/\sqrt{24} = 2.2706 \qquad [3.2]$$

$$\text{Lower Limit} = e^{L(\bar{x})} = 9.7'$$

Clearly, 9.7 mg/kg is greater than 5 mg/kg and we reject H_0.

Obviously, each of the above decision rules has led to the rejection of H_0. In doing so we can only make an error of Type I and the probability of making such an error has been fixed at 5% ($\alpha = 0.05$). Let us say that the remediation of our residential exposure unit will cost $1 million. A 5% chance of error in the decision to remediate results in an expected loss of $50,000. That is simply the cost to remediate, $1 million, times the probability that the decision to remediate is wrong ($\alpha = 0.05$). However, the calculated value of the "t" statistic, t = 4.84, is well above the 95th percentile of the "t"-distribution.

We might ask exactly what is the probability that a value of t equal to or greater than 4.84 will result when H_0 is true. This probability, "P," can be obtained from tables of the student's "t"-distribution or computer algorithms for computing the cumulative probability function of the "t"-distribution. The *"P" value* for the current example is 0.00003. Therefore, the expected loss in deciding to remediate this particular exposure unit is likely only $30.

There is another use of the *"P" value*. Instead of comparing the calculated value of the test statistic to the tabulated value corresponding to the Type I error probability to make the decision to reject H_0, we may compare the "P" value to the tolerable Type I error probability. If the "P" value is less than the tolerable Type I error probability we then will reject H_0.

Test Operating Characteristic

We have now considered the ramifications associated with the making of a Type I decision error, i.e., rejecting H_0 when it is in fact true. In our example we are 95% confident that the true median concentration is greater than 9.7 mg/kg and it is therefore unlikely that we would ever get a sample from our remedial unit that would result in accepting H_0. However, this is only a *post hoc* assessment. Prior to collecting the statistical collection of physical soil samples from our exposure unit it seems prudent to consider the risk making a *false negative decision error*, or error of Type II.

Unlike the probability of making a Type I error, which is neither a function of the sample size nor the true deviation from H_0, the probability of making a Type II error is a function of both. Taking the effect of the deviation from a target median of 5 mg/kg and the sample size separately, let us consider their effects on the probability, β, of making a Type II error.

Figure 3.1 presents the probability of a Type II error as a function of the true median for a sample size of 24. This representation is often referred to as the *operating characteristic* of the test. Note that the closer the true median is to the target value of 5 mg/kg, the more likely we are to make a Type II decision error and accept H_0 when it is false. When the true median is near 14, it is extremely unlikely that will make this decision error.

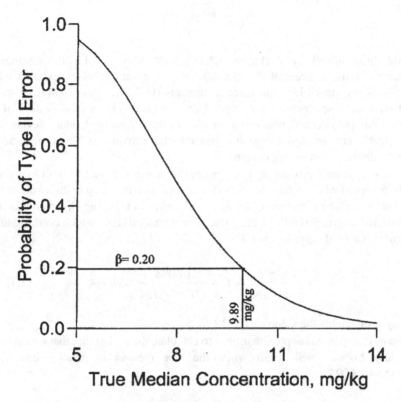

Figure 3.1 **Operating Characteristic,**
Single Sample Student's t-Test

It is not uncommon to find a false negative error rate specified as 20% ($\beta = 0.20$). The choice of the tolerable magnitude of a Type II error depends upon the consequent costs associated with accepting H_0 when it is in fact false. The debate as to precisely what these costs might include, i.e., remobilization and remediation, health care costs, cost of mortality, are well beyond the scope of this book. For now we will assume that $\beta = 0.20$ is tolerable.

Note from Figure 3.1 that for our example, a $\beta = 0.20$ translates into a true median of 9.89 mg/kg. The region between a median of 5 mg/kg and 9.89 mg/kg is often referred to as the "gray area" in many USEPA guidance documents (see for example, USEPA, 1989, 1994a, 1994b). This is the range of the true median greater than 5 mg/kg where the probability of falsely accepting the null hypothesis exceeds the tolerable level. As is discussed below, the extent of the gray region is a function of the sample size.

The calculation of the exact value of β for the student's "t"-test requires the evaluation of the noncentral "t"-Distribution with noncentrality parameter d, where d is given by

$$d = \frac{\sqrt{N}(\mu - \mu_a)}{\sigma}$$

Several statistical software packages such as SAS® and SYSTAT® offer routines for evaluation of the noncentral "t"-distribution. In addition, tables exist in many statistical texts and USEPA guidance documents (USEPA, 1989, 1994a, 1994b) to assist with the assessment of the Type II error. All require a specification of the noncentrality parameter d, which is a function of the unknown standard deviation σ. A reasonably simple approximation is possible that provides sufficient accuracy to evaluate alternative sampling designs.

This approximation is simply to calculate the probability that the null hypothesis will be accepted when in fact the alternate is true. The first step in this process is to calculate the value of the mean, $\bar{x}$, which will result in rejecting H_0 when it is true. As indicated above, this will be the value of $\bar{x}$, let us call it C, which corresponds to the critical value of $t_{v=23,\,0.95} = 1.714$:

$$t = \frac{C - \mu_0}{S/\sqrt{N}} = C - \frac{1.6094}{1.0357/\sqrt{24}} = 1.714 \qquad [3.3]$$

Solving for C yields the value of 1.9718.

The next step in this approximation is to calculate the probability that a value of $\bar{x}$ less than 2.06623 will result when the true median is greater than 5, or $\mu > \ln(5) = 1.6094$:

$$\Pr(\bar{x} < C | \mu > \mu_a) = \beta$$

$$\Pr(\bar{x} < 1.9718 | \mu > 1.6094) = \beta \qquad [3.4]$$

Suppose that a median of 10 mg/kg is of particular interest. We may employ [3.4] with $\mu = \ln(10) = 2.3026$ to calculate β:

$$\beta = \Pr\left(t \le \frac{C - \mu}{S/\sqrt{N}} = \frac{1.9718 - 2.3026}{0.2114} = -1.5648\right)$$

Using tables of the Student's "t"-distribution, we find $\beta = 0.066$, or, a Type II error rate of about 7%.

Power Calculation and One Sample Tests

A function often mentioned is referred to as the discriminatory power, or simply the **power**, of the test. It is simply one minus the magnitude of the Type II error, or power $= 1-\beta$. The power function for our example is presented in Figure 3.2. Note that there is at least an 80 percent chance of detecting a true median as large as 9.89 mg/kg and declaring it statistically significantly different from 5 mg/kg.

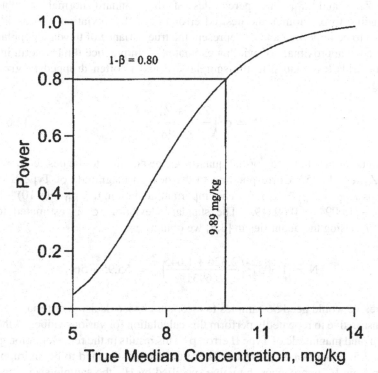

Figure 3.2   Power Function,
Single Sample Student's t-Test

Sample Size

We discovered that there is a 14 percent chance of accepting the hypothesis that the median concentration is less than or equal to 5 mg/kg when in truth the median is as high as 10 mg/kg. There are situations in which a doubling of the median concentration dramatically increases the consequences of exposure. Suppose that this is one of those cases. How can we modify the sampling design to reduce the magnitude of the Type II error to a more acceptable level of $\beta = 0.01$ when the true median is 10 ($\mu = \ln(10) = 2.3026$)?

Step 7 of the DQO process addresses precisely this question. It is here that we combined our choices for magnitudes α and β of the possible decision errors, an estimate of the data variability with perceived important deviation of the mean from that specified in H_0 to determine the number of samples required. Determining the exact number of samples requires iterative evaluation of the probabilities of the noncentral t distribution. Fortunately, the following provides an adequate approximation:

$$N = \sigma^2\left(\frac{Z_{1-\beta} + Z_{1-\alpha}}{\mu - \mu_0}\right)^2 + \frac{Z_{1-\alpha}^2}{2} \qquad [3.5]$$

Here $Z_{1\alpha}$ and $Z_{1\beta}$ are percentiles of the standard normal distribution corresponding to one minus the desired error rate. The deviation $\mu - \mu_0$ is that considered to be important and σ^2 represent the true variance of the data population. In practice we approximate σ^2 with an estimate S^2. In practice the last term in this expression adds less than 2 to the sample size and is often dropped to give the following:

$$N = \sigma^2 \left(\frac{Z_{1-\beta} + Z_{1-\alpha}}{\mu - \mu_0} \right)^2 \qquad [3.6]$$

The value of the standard normal quantile corresponding to the desired $\alpha = 0.05$ is $Z_{1\alpha}$ $Z_{0.95} = 1.645$. Corresponding to the desired magnitude of Type II error, $\beta = 0.01$, is $Z_{1\beta} = Z_{0.99} = 2.326$. The important deviation, $\mu - \mu_0 = \ln(10) - \ln(5) = 2.3026 - 1.6094 = 0.69319$. The standard deviation, σ, is estimated to be $S = 1.3057$. Using the quantities in [3.6] we obtain

$$N = 1.3057^2 \left(\frac{2.326 + 1.645}{0.69319} \right)^2 = 55.95 \approx 56$$

Therefore, we would need 56 samples to meet our chosen decision criteria.

It is instructive to repeatedly perform this calculation for various values of the log median, μ, and magnitude of Type II error, β. This results in the representation given in Figure 3.3. Note that as the true value of the median deemed to be an important deviation from H_0 approaches the value specified by H_0, the sample size increases dramatically for a given Type II error. Note also that the number of samples also increases as the tolerable level of Type II error decreases.

Frequently, contracts for environmental investigations are awarded based upon minimum proposed cost. These costs are largely related to the number of samples to be collected. In the authors' experience candidate project proposals are often prepared without going through anything approximating the steps of the DQO process. Sample sizes are decided more on the demands of competitive contract bidding than analysis of the decision making process. Rarely is there an assessment of the risks of making decision errors and associated economic consequences.

The USEPA's *Data Quality Objects Decision Error Feasibility Trails, (DQO/DEFT)* program and guidance (USEPA 1994c) provides a convenient and potentially useful tool for the evaluation of tolerable errors alternative sampling designs. This tool assumes that the normal theory of errors applies. If the normal distribution is not a useful model for hypothesis testing, this evaluation requires other tools.

Whose Ox is Being Gored

The astute reader may have noticed that all of the possible null hypotheses given above specify the unit sampled as being "clean." The responsible party therefore has a fixed specified risk, the Type I error, that a "clean" unit will be judged "contaminated" or a discharge in compliance as noncompliant. This is not always the case.

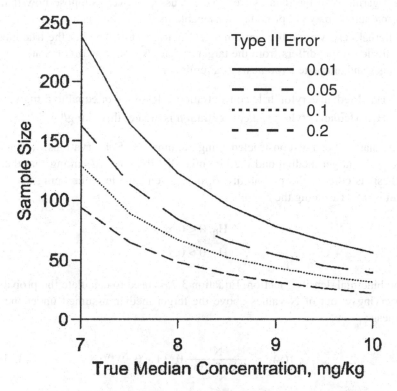

Figure 3.3 **Sample Sizes versus True Median Concentration**
for Various Type II Errors
(Type I Error Fixed at $\alpha = 0.05$)

The USEPA's (1989) *Statistical Methods for Evaluating the Attainment of Cleanup Standards, Volume 1: Soils and Solid Media,* clearly indicates that "it is extremely important to say that the site shall be cleaned up until the sampling program indicates with reasonable confidence that the concentrations of the contaminants at the entire site are statistically less than the cleanup standard" (USEPA 1994a, pp. 2–5). The null hypothesis now changes to "the site remains contaminated until proven otherwise within the bounds of statistical certainty." The fixed Type I error is now enjoyed by the regulating parties. The responsible party must now come to grips with the "floating" risk, Type II error, of a truly remediated site being declared contaminated and how much "overremediation" is required to control those risks.

Nonparametric Tests

We thus far have assumed that a lognormal model provided a reasonable model for our data. The geometric mean and median are asymptotically equivalent for the lognormal distribution, so a test of median is in effect a test geometric mean or mean of the logarithms of the data as we have discussed above. Suppose now that the lognormal model may not provide a reasonable model for our data.

Alternatively, we might want a nonparametric test of whether the true median acrylonitrile sample differs from the target of 5 mg/kg. Let us first restate our null hypothesis and alternate hypothesis as a reminder:

H_0: Median acrylonitrile concentration is less than or equal to 5 mg/kg;
H_1: Median acrylonitrile concentration is greater than 5 mg/kg.

A median test can be constructed using the number of observations, w, found to be above the target median and the binomial distribution. Assuming that the null hypothesis is correct, the probability, θ, of a given sample value being above the median is 0.5. Restating the hypothesis:

$$H_0, \theta \leq 0.5$$

$$H_1, \theta > 0.5$$

The binomial density function, Equation 3.7, is used to calculate the probability of observing w out of N values above the target median assumed under the null hypothesis:

$$f(w) = \frac{N!}{w!(N-w)!}\theta^w(1-\theta)^{N-w} \qquad [3.7]$$

To test H_0 with a Type I error rate of 5% ($\alpha = 0.05$), we find a critical value, C, as the largest integer that satisfies the inequality:

$$Pr(w < C | \theta \leq 0.5) = \sum_{w=0}^{C-1} f(w) \leq (1-\alpha) = 0.95 \qquad [3.8]$$

If we observe C or more values greater than our assumed background, we then reject H_0. For our example, C is 17 and we observe k = 20 values greater than background; thus we reject H_0. Note that if we want to determine the probability, "P-value," of observing w or more successes, where k is the observed number above the median (20 in our example), we sum f(w) from w = k to N. For our example, the P-value is about 0.0008.

We can also assess the Type II error by evaluating Equation [3.8] for values of $\theta > 0.5$:

$$Pr(w < C | \theta > 0.5) = \sum_{w=0}^{C-1} f(w) = \beta \qquad [3.9]$$

The following Table 3.3 presents the magnitude of the Type II error for our current example for several values of θ greater than 0.5.

Table 3.3
Probability of Type II Error versus θ > 0.5

θ	β
0.55	0.91
0.60	0.81
0.65	0.64
0.70	0.44
0.75	0.23
0.80	0.09

Tests Involving Two Samples

Rather than comparing the mean or median of a single sample to some fixed level, we might wish to consider a question like: "Given that we have sampled 18 observations each from two areas, and have obtained sample means of 10 and 12 ppm, what is the probability that these areas have the same population mean?" We could even ask the question "If the mean concentration of bad stuff in areas A and B differs by 5 ppm, how many samples do we have to take from areas A and B to be quite sure that the observed difference is real?"

If it can be assumed that the data are reasonably represented by the normal distribution model (or if the logarithms represented by a normal distribution; e.g., log-normal) we can use the same t-test as described above, but now our population mean is $\mu_1 - \mu_2$; that is, the difference between the two means of the areas of interest. Under the null hypothesis the value of $\mu_1 - \mu_2$ is zero and $\bar{x}_1 - \bar{x}_2$ has a "t"-distribution. The standard deviation used for this distribution is derived from a "pooled" variance, S_p^2, given by:

$$S_p^2 = \frac{(N_1 - 1) S_1^2 + (N_2 - 1) S_2^2}{N_1 + N_2 - 2}$$ [3.10]

This pooled variance is taken as the best overall estimate of the variance in the two populations if we assume that the two populations have equal variances.

Once we have calculated S_p^2, we can use the principal that the variance of the difference of two random variables is the sum of their variances (Hogg and Craig, 1995). In our case the variance of interest is the variance of $\bar{x}_1 - \bar{x}_2$, which we will call S_D^2. Since we know that the variance of the sample mean is given by S^2/N

(Equation [2.27]), it follows that the variance of the difference between two sample means, S_D^2 (assuming equal variances) is given by:

$$S_D^2 = S_p^2\left(\frac{1}{N_1} + \frac{1}{N_2}\right) \qquad [3.11]$$

and the standard deviation of the difference is its square root, S_D.

The 95% confidence interval for $\bar{x}_1 - \bar{x}_2$ is defined by an upper confidence bound, $U_{\bar{x}1 - \bar{x}2}$ for a two-sided probability interval of width $(1\alpha\ )$, given by:

$$U_{\bar{x}1 - \bar{x}2} = \bar{x}_1 - \bar{x}_2 + t_{v1 + v2,(1 - \alpha/2)} S_D \qquad [3.12]$$

and a lower confidence bound, $L_{\bar{x}1 - \bar{x}2}$ or a two-sided probability interval of width $(1\alpha\ )$, given by:

$$L_{\bar{x}1 - \bar{x}2} = \bar{x}_1 - \bar{x}_2 - t_{v1 + v2,(1 - \alpha/2)} S_D \qquad [3.13]$$

If we were doing a two-sided hypothesis with an alternative hypothesis H_1 of the form $\bar{x}_1$ and $\bar{x}_2$ are not equal, we would reject H_0 if the interval $(L_{\bar{x}1 - \bar{x}2}, U_{\bar{x}1 - \bar{x}2})$ does not include zero.

One can also pose a one-tailed hypothesis test with an alternate hypothesis of the form $\bar{x}_1$ is greater than $\bar{x}_2$. Here we would reject H_0 if

$$L_{\bar{x}1 - \bar{x}2} = \bar{x}_1 - \bar{x}_2 - t_{v1 + v2,(1 - \alpha)} S_D \qquad [3.14]$$

were less than zero (note that for the one-tailed test we switch from $\alpha/2$ to α).

One point that deserves further consideration is that we assumed that S_1^2 and S_2^2 were equal. This is actually a testable hypothesis. If we have S_1^2, S_2^2 and want to determine whether they are equal, we simply pick the larger of the two variances and calculate their ratio, F, with the larger as the numerator. That is, if S_1^2 were larger than S_2^2, we would have:

$$F = S_1^2/S_2^2 \qquad [3.15]$$

This is compared to the critical value of an F distribution with $(N_1 - 1)$ and $(N_2 - 1)$ degrees of freedom, which is written as $F_{\alpha/2[v1,v2]}$. Note that the actual test has

$$H_0: S_1^2 = S_2^2, \text{ and}$$

$$H_1: S_1^2 \neq S_2^2$$

that is, it is a two-tailed test, thus we always pick the larger of S_1^2 and S_2^2 and test at a significance level of $\alpha/2$. For example, if we wanted to test equality of variance at a significance level of 0.05, and we have sample sizes of 11 and 12, and the larger variance was from the sample of size 12, we would test S_{12}^2/S_{11}^2 against $F_{0.025\ [11,10]}$ (remember degrees of freedom for the sample variance is always $N-1$).

We note that many statistics texts discuss modifications of the t-test, generally referred to as a Behrens-Fisher t-test, or Behrens-Fisher test, or a Behrens-Fisher correction for use when sample variances are unequal (e.g., Sokol and Rohlf, 1995; Zar, 1996). It is our experience that when unequal variances are encountered, one should first try a logarithmic transformation of the data. If this fails to equalize variances, one should then consider the nonparametric alternative discussed below, or if differences in arithmetic means are the focus of interest use bootstrap methods (Chapter 6). The reason for our not recommending Behrens-Fisher t-tests is that we have seen such methods yield quite poor results in real-world situations and feel that rank-based or bootstrap alternatives are more robust.

The following example uses the data from Table 2.4 to illustrate a two-sample t-test and equality-of-variance test. The values from the two samples are designated by "sample ID" in column 1 of Table 2.4. The summary statistics required for the conduct of the hypothesis test comparing the means of the two populations are as follows:

Sample No. 1:

$$\bar{x}_1 = 2.6281$$

$$S_1^2 = 0.8052$$

$$N_1 = 15.$$

Sample No. 2:

$$\bar{x}_2 = 4.0665$$

$$S_2^2 = 0.5665$$

$$N_2 = 15.$$

The first hypothesis to be considered is the equality of variances:

$$F = S_1^2/S_2^2 = 0.8052/0.5665 = 1.421$$

The critical value of $F_{0.025,\ [14,14]} = 2.98$. Since $F = 1.421$ is less than the critical value of 2.98, there is no indication of unequal variances. Therefore, we may calculate the pooled variance using Equation [3.10] and $S_p^2 = 0.68585$. Consequently, the standard deviation of the difference in the two means is $S_D = 0.3024$ using Equation [3.11]. Employing relationships [3.12] and [3.13] we obtain the 95% confidence interval for the true mean difference as $(-2.0577, -0.8191)$. Because this interval does not contain zero, we reject the null hypothesis H_0.

One thing that may strike the careful reader is that in Chapter 2 we decided that the data were consistent with a normal distribution, yet when we do a t-test we declare that the two samples have significantly different means. This may seem contradictory, but it is not; the answer one gets from a statistical test depends on the question one asks.

In Chapter 2 we asked, "Are the data consistent with a normal distribution?" and received an affirmative answer, while here we have asked, "Do the two samples have the same mean?" and received a negative answer. This is actually a general principle. One may have a population that has an overall distribution that is well described by a single distribution, but at the same time have subpopulations that are significantly different from one another. For example, the variation in height of male humans can be well described by a normal distribution, but different male populations such as jockeys and basketball players may have very different mean heights.

Power Calculations for the Two-Sample t-Test

Determination of the power of the two-sample test is very similar to that of the sample test; that is, under H_0, $\mu_1 - \mu_2$ is always assumed to be zero. If under H_1 we assume that $\mu_1 - \mu_2 = \delta$, we can determine the probability that we will reject H_0 when it is false, which is the power of the test. The critical value of the test is $t_{v1+v2, (1\alpha\ /2)}$ S_D or $-t_{v1+v2, (1\alpha\ /2)}$ S_D because our expected mean difference is zero under H_0. If we consider an H_1 of $\mu_1 < \mu_2$ with a mean difference of δ, we want to calculate the probability that a distribution with a true mean of δ will yield a value greater than the upper critical value $C_L = -t_{v1+v2, (1\alpha\ /2)}$ S_D (we are only interested in the lower bound because H_1 says $\mu_1 - \mu_2 < \delta$). In this case, we obtain a $t_{v1+v2,(\beta)}$ as:

$$t_{v1+v2,(\beta)} = (\delta - C_L) / S_D \qquad [3.16]$$

We then determine the probability of a t statistic with $v1 + v2$ degrees of freedom being greater than the value calculated using [3.17]. This is the power of the t-test. We can also calculate sample sizes required to achieve a given power for a test with a given α level. If we assume that our two sample sizes will be equal (that is, $N_1 = N_2 = N$), we can calculate our required N for *each* sample as follows:

$$N = (2S_p^2/\delta^2) (t_{v\alpha} + t_{v\beta}) \qquad [3.17]$$

Here $t_{v(\alpha)} + t_{v(\beta)}$ are the t values associated with the α level of the test ($\alpha/2$ for a two-tailed test) and S_p^2 and δ are as defined above.

The observant reader will note that v is given by $2N - 2$, but we are using [3.17] to calculate N. In practice this means we must take a guess at N and then use the results of the guess to fine tune our N estimate. Since N is usually fairly large, one good way to get an initial estimate is to use the normal statistics, Z_α and Z_β to get an initial N estimate, and then use this N to calculate v for our t distribution. Since $t_{v(\alpha)}$ and $t_{v(\beta)}$ will always be slightly larger than Z_α and Z_β (see Table 2.2), our initial N will always be a little too small. However, in general, a sample size one or two units higher than our initial N guess will usually satisfy [3.17]. One can also do more complex power calculations where N_1 might be a fixed multiple of N_2. Such a design may be desirable if samples from population 1 are less expensive to obtain than samples from population 2. More extensive discussions of power calculations for t-tests can be found in Sokol and Rohlf (1995) and Zar (1996).

A Rank-Based Alternative to the Two-Sample t-Test

In the previous section, we performed the two-sample t-test, but if the data are not from a normal distribution or the variances of the two samples are not equal, the probability levels calculated may be incorrect. Therefore, we consider a test alternative that does not depend on assumptions of normality or equality of variance. If we simply rank all of the observations in the two samples from smallest to largest and sum the ranks of the observations in each sample, we can calculate what is called the Mann Whitney U test or Wilcoxon Rank Sum Test (Conover, 1998; Lehmann, 1998).

The U statistic is given by:

$$U = (N_1 N_2) + (N_1(N_1 + 1)/2) - R_1 \qquad [3.18]$$

Here N_1 and N_2 are the sizes of the two samples and R_1 is the sum of the ranks in sample 1. One might ask, "How do I determine which sample is sample 1?" The answer is that it is arbitrary and one must calculate U values for both samples. However, once a U value has been determined for one sample, a U′ value that would correspond to the other sample can easily be determined as:

$$U' = (N_1 N_2) - U \qquad [3.19]$$

Using our two-sample example from Table 2.4, we obtain the following:

Sample	Size N	Rank Sum R
No. 1	15	144
No. 2	15	321
Total	30	465

Using [3.18] and [3.19] we obtain U = 201 and U′ = 24, and compare the *smaller* of the two values to a table like that in Table 3.4. If this value is *less* than the tabulated critical value we reject H_0 that the sampled populations are the same. U′ = 24 is certainly less than the tabulated 72, so we have two different populations sampled in our example. Note that one can base the test on either the larger or the smaller of the U values. Thus, when using other tables of critical values, it is important to determine which U (larger or smaller) is tabulated.

In practice, statistical software will always provide P values for the U statistics. If one has a fairly large sample size (as a rule of thumb: N1 + N2 ≥ 30 and the smaller of the two sample sizes greater than 10), one can calculate an average U value, U_M, as:

$$U_M = (N_1 N_2)/2 \qquad [3.20]$$

and a standard error for U, S_U as:

$$S_U = [(N_1 N_2) \ (N_1 + N_2 + 1) \ / 12]^{1/2} \qquad [3.21]$$

Table 3.4
Critical Values of U in the Mann-Whitney Test
($\alpha = 0.05$ for a One-Tailed Test, $\alpha = 0.10$ for a Two-Tailed Test)

N_2	N_1											
	9	10	11	12	13	14	15	16	17	18	19	20
1											0	0
2	1	1	1	2	2	2	3	3	3	4	4	4
3	3	4	5	5	6	7	7	8	9	9	10	11
4	6	7	8	9	10	11	12	14	15	16	17	18
5	9	11	12	13	15	16	18	19	20	22	23	25
6	12	14	16	17	19	21	23	25	26	28	30	32
7	15	17	19	21	23	26	28	30	33	35	37	39
8	18	20	23	26	28	31	33	36	39	41	44	47
9	21	24	27	30	33	36	39	42	45	48	51	54
10	24	27	31	34	37	41	44	48	51	55	58	62
11	27	31	34	38	42	46	50	54	57	61	65	69
12	30	34	38	42	47	51	55	60	64	68	72	77
13	33	37	42	47	51	56	61	65	70	75	80	84
14	36	41	46	51	56	61	66	71	77	82	87	92
15	39	44	50	55	61	66	72	77	83	88	94	100
16	42	48	54	60	65	71	77	83	89	95	101	107
17	45	51	57	64	70	77	83	89	96	102	109	115
18	48	55	61	68	75	82	88	95	102	109	116	123
19	51	58	65	72	80	87	94	101	109	116	123	130
20	54	62	69	77	84	92	100	107	115	123	130	138

Adapted from *Handbook of Tables for Probability and Statistics*, CRC Press.

The Z score is then

$$Z = (U - U_M) / S_U \qquad [3.22]$$

The result of Equation [3.22] is then compared to a standard normal distribution, and H_0 is rejected if Z is greater than $Z_{(1\alpha\ /2)}$. That is, if we wished to do a two-sided hypothesis test for H_0 we would reject H_0 if Z exceeded 1.96.

One question that arises is "exactly what is H_0?" For the t-test it is $\mu_1 = \mu_2$, but for a rank sum test H_0, is that the ranks are assigned randomly to the two samples, which is essentially equivalent to an H_0 that the two sample medians are equal. In some cases, such as sampling for exposure assessment, we may be specifically interested in H_0: $\mu_1 - \mu_2$, where $\bar{x}_1$ and $\bar{x}_2$ are the sample arithmetic means. For strongly right-skewed distributions, such as the log-normal-like ones associated with chemical concentration data, the arithmetic mean may be the 75th or even 90th percentile of the distribution. Thus a test of medians may be misleading. In such cases, tests based on bootstrapping are a better alternative.

Another problem with rank tests is tied values. That is, one may have two observations with the same value. This may occur in environmental measurements because reported values are rounded to a small number of decimal places. If the number of ties is small, one can simply assign the average rank to each of the tied values. That is, if two values that are tied at the positions that would ordinarily be assigned ranks 7 and 8, each is assigned 7.5. One then simply calculates U and U' and ignores the ties when doing the hypothesis test. In this case the test is slightly conservative in the sense that it is less likely to reject the null hypothesis than if we calculated an exact probability (which could always be done using simulation techniques). Lehmann (1998) discusses the problem of ties and most discussions of this test (e.g., Conover, 1998) offer formulae for large sample corrections for ties. It is our feeling that for these cases, too, bootstrap alternatives are preferable.

A Simple Two-Sample Quantile Test

Sometimes we are not totally interested in the mean values but rather want to determine if one area has more "high" concentration values than another. For example, we might want to know if a newly remediated area has no more spot contamination than a "clean" reference area. In this case we might simply pick some upper quantile of interest such as the upper 70th or 80th percentile of the data and ask whether the remediated area had more observations greater than this quantile than the reference area.

Let us again consider the data in Table 3.4. Suppose that the data of sample No. 1 come from an acknowledged reference area. Those data identified as from sample No. 2 are from an area possibly in need of remediation. It will be decided that the area of interest has no more "high" concentration values than the reference area if it is statistically demonstrated that the number of observations from each area greater than the 70th percentile of the combined set of values is the same. Further, we will fix our Type I error at $\alpha = 0.05$. The exact P-value of the quantile test can be obtained from the hypergeometric distribution as follows:

$$P = \frac{\sum\limits_{i=k}^{r} \binom{m+n-r}{n-i}\binom{r}{i}}{\binom{m+n}{n}} \qquad [3.23]$$

We start by sorting all the observations from the combined samples and note the upper 70th percentile. In our example, this is $\ln(59.8) = 4.09123$. Let r (=9) be the total number of observations above this upper quantile. The number of observations from the area of interest greater than or equal to this value is designated by k (=8). The total number of samples from the reference area will be represented by m (=15) and the total number of samples from the area of interest by n (=15):

$$P = \frac{\sum\limits_{i=8}^{9} \binom{21}{15-i}\binom{9}{i}}{\binom{30}{15}} = 0.007$$

Thus, we reject the hypothesis that the area of interest and the reference area have the same frequency of "high" concentrations.

If the total number of observations above the specified quantile, r, is greater than 20, the calculation of the hypergeometric distribution can become quite tedious. We may then employ the approximation involving the normal distribution. We first calculate the mean, μ, and standard deviation, σ, of the hypergeometric distribution assuming H_0 is true:

$$\mu = \frac{nr}{m+n} \qquad [3.24]$$

$$\sigma = \left(\frac{mnr(m+n-r)}{(m+n)^2(m+n-1)}\right)^{\frac{1}{2}} \qquad [3.25]$$

The probability used to determine significance is that associated with the standard normal variate Z found by:

$$Z = \frac{k - 0.5 - \mu}{\sigma} \qquad [3.26]$$

The Quantile Test is a prominent component in making decisions regarding the success of site cleanups. It is a major part of the USEPA's (1994a) *Statistical Methods For Evaluating The Attainment of Cleanup Standards* for soils and solid media and the NRC's (1995) NUREG-1505 on determining the final status of decommissioning surveys. These documents recommend that the Quantile Test be used in conjunction with the Wilcoxon Rank Sum Test.

More Than Two Populations: Analysis of Variance (ANOVA)

In some cases we may have several samples and want to ask the question, "Do these samples have the same mean?" (H_0) or "Do some of the means differ?" (H_1). For example we might have a site with several distinct areas and want to know if it is reasonable to assume that all areas have a common mean concentration for a particular compound.

To answer such a question we do a one-way ANOVA of the replicate x data across the levels of samples of interest. In such a test we first calculate a total sum of squares (SS_T) for the data set, which is given by:

$$SS_T = \sum_{i=1}^{M} \sum_{j=1}^{K_i} (x_{i,j} - \bar{x}_G)^2 \qquad [3.27]$$

where $\bar{x}_G$ is the grand mean of the x's from all samples. M is the number of samples of interest and K_i is the sample size in the ith group.

We then calculate a within-group sum of squares, SS_W, for each group. This is given by:

$$SS_W = \sum_{i=1}^{M} \sum_{j=1}^{K_i} (x_{i,j} - \bar{x}_{i.})^2 \qquad [3.28]$$

Here, K_i and M are defined as before; $\bar{x}_{i.}$ is the mean value for each group.

We can then calculate a between-group sum of squares (SS_B) by subtraction:

$$SS_B = SS_T - SS_W \qquad [3.29]$$

Once we have calculated SS_W and SS_B, we can calculate "mean square" estimates for within- and between group variation (MS_W and MS_B):

$$MS_W = SS_W / \sum_{i=1}^{M} (K_i - 1) \text{ , and } MS_B = SS_B/(N-1) \qquad [3.30]$$

These are actually variance estimates. Thus, we can test whether MS_B and MS_W are equal using an F test like that used for testing equality of two sample variances, except here:

$$H_0 \text{ is } MS_B = MS_W, \text{ versus } H_1, MS_B > MS_W$$

These hypotheses are equivalent to H_0 of "all means are equal" versus an H_1 of some means are unequal because when all means are equal, both MS_B and MS_W are estimates of the population variance, σ^2 and when there are differences among means, MS_B is larger than MS_W. We test the ratio:

$$F = MS_B / MS_W \qquad\qquad [3.31]$$

This is compared to the critical value of an F distribution with $(N-1)$ and $\Sigma(K_i - 1)$ degrees of freedom, which is written as: $F_{\alpha[v1,v2]}$. Note that here we test at a level α rather than $\alpha/2$ because the test is a one-tailed test. That is, under H_1, MS_B is always greater than MS_W.

Assumptions Necessary for ANOVA

There are two assumptions necessary for Equation [3.31] to be a valid hypothesis test in the sense that the α level of the test is correct. First, the data must be normally distributed and second, the M groups must have the same variance. The first assumption can be tested by subtracting the group mean from the observations in each group. That is, $x_{i,j,C}$ is found as:

$$x_{i,j,C} = x_{i,j} - \bar{x}_i \qquad\qquad [3.32]$$

The N ($N = \Sigma K_i$) total $x_{i,j,C}$ values are then tested for normality using either the Kolmogorov-Smirnov test or the correlation coefficient between the $x_{i,j,C}$ and their expected normal scores as described in Chapter 2.

The most commonly used test for equality of variances is Bartlett's test for homogeneity of variances (Sokol and Rohlf, 1995). For this test we begin with the MS_W value calculated in our ANOVA and the variances of each of the M samples in the ANOVA, $S_1^2, \ldots, S_M^2$. We then take the natural logs of the MS_W and the M within-sample S^2 values. We will write these as L_W and $L_1, \ldots L_M$. We develop a test statistic, χ^2 as:

$$\chi^2 = C\left[L_W \sum_{i=1}^{M} (K_i - 1) - \sum_{i=1}^{M} L_i(K_i - 1) \right] \qquad\qquad [3.33]$$

This is compared to a chi-squared statistic with $M-1$ degrees of freedom.

In Equation [3.33], C is given by:

$$C = 1 + A(B - D)$$

where

$$A = 1/3(M-1) \quad B = \sum_{i=1}^{M} 1/(K_i - 1) \quad D = 1/\sum_{i=1}^{M} (K_i - 1) \qquad [3.34]$$

Table 3.5 provides a sample one-way ANOVA table. The calculations use the log-transformed pesticide residue data, x, found in Table 3.6. Table 3.6 also provides the data with the group means (daily means) subtracted. The F statistic for this analysis has 8 and 18 degrees of freedom because there are 9 samples with 3 observations per sample. Here the log-transformed data are clearly normal (the interested reader can verify this fact), and the variances are homogeneous (the Bartlett χ^2 is not significant). The very large F value of 92.1 is highly significant (the P value of 0.0000 means that the probability of an F with 8 and 18 degrees of freedom having a value of 92.1 or more is less than 0.00001).

Table 3.5
ANOVA Pesticide Residue Example

Source of Variation	Degrees of Freedom	Sum of Squares	Mean Square	F Statistic	P Value
Days	8	98.422	12.303	92.1	<0.00001
Error	18	2.405	0.134		
Total	26	100.827			

Table 3.6
Data for Pesticide Example with Residuals and Ranks

Day	Residual Pesticide, y, (ppb)	x = ln(y)	Deviation from Daily Mean, $X - \bar{X}$	Rank Order	Group Mean Rank
0	239	5.4764	−0.11914	20.0	
0	232	5.4467	−0.14887	19.0	20.8
0	352	5.8636	0.26802	23.5	
1	256	5.5452	0.13661	21.0	
1	116	4.7536	−0.65497	16.0	21.0
1	375	5.9269	0.51836	26.0	
5	353	5.8665	−0.14014	25.0	
5	539	6.2897	0.28311	27.0	25.2
5	352	5.8636	−0.14297	23.5	
10	140	4.9416	−0.36377	17.0	
10	269	5.5947	0.28929	22.0	19.0
10	217	5.3799	0.07448	18.0	
20	6	1.7664	0.06520	8.0	
20	5	1.5063	−0.19494	6.0	8.0

Table 3.6 (Cont'd)
Data for Pesticide Example with Residuals and Ranks

Day	Residual Pesticide, y, (ppb)	x = ln(y)	Deviation from Daily Mean, $X - \overline{X}$	Rank Order	Group Mean Rank
20	6	1.8310	0.12974	10.0	
30	4	1.4303	0.02598	3.0	
30	4	1.4770	0.07272	5.0	3.0
30	4	1.3056	−0.09870	1.0	
50	4	1.4702	−0.24608	4.0	
50	5	1.6677	−0.04855	7.0	7.7
50	7	2.0109	0.29464	12.0	
70	8	2.0528	0.03013	13.0	
70	4	1.3481	−0.67464	2.0	10.0
70	14	2.6672	0.64451	15.0	
140	6	1.7783	−0.22105	9.0	
140	7	1.9242	−0.07513	11.0	11.3
140	10	2.2956	0.29617	17.0	

Power Calculations for ANOVA

One can calculate the power for an ANOVA in much the same way that one does them for a t-test, but things get very much more complex. Recall that the H_0 in ANOVA is that "all means are the same" versus an H_1 of "some means are different." However for the power calculation we must have an H_1 that is stated in a numerically specific way. Thus we might have an H_1 that all means are the same except for one that differs from the others by an amount δ. Alternatively, we might simply say that the among-group variance component exceeded the within group component by an assumed amount.

It is our feeling that power or sample size calculations for more complex multisample experimental designs are best pursued in collaboration with a person trained in statistics. Thus, we do not treat such calculations here. Those wishing to learn about such calculations can consult Sokol and Rohlf (1995; Chapter 9) or Zar (1996; Chapters 10 and 12). For a more extensive discussion of ANOVA power calculations one can consult Brown et al. (1991).

Multiway ANOVA

The preceding discussion assumed a group of samples arrayed along a single indicator variable (days in our example). Sometimes we may have groups of samples defined by more than one indicator. For example, if we had collected pesticide residue data from several fields we would have samples defined by days and fields. This would be termed a two-way ANOVA. Similarly, if we had a still larger data set that represented residues collected across days, and fields and several years we would have a three-way ANOVA.

In our experience, multiway ANOVAs are not commonly employed in environmental quality investigations. However, we mention these more complex analyses so that the reader will be aware of these tools. Those desiring an accessible account of multiway ANOVAs should consult Sokol and Rohlf (1995; Chapter 12) or Zar (1996; Chapters 14 and 15). For a more comprehensive, but still relatively nonmathematical, account of ANOVA modeling we suggest Brown et al. (1991).

A Nonparametric Alternative to a One-Way ANOVA

Sometimes either the data do not appear to be normal and/or the variances are not equal among groups. In such cases the alternative analysis is to consider the ranks of the data rather than the data themselves. The procedure of choice is the Kruskal-Wallis test (Kruskal and Wallis, 1952; Zar, 1996). In this test all of the data are ranked smallest to largest, and the ranks of the data are used in the ANOVA.

If one or more observations are tied, all of the tied observations are assigned the average rank for the tied set. That is, if 3 observations share the same value and they would have received ranks 9, 10, and 11, all three receive the average rank, 10. After the ranks are calculated we sum the ranks separately for each sample. For example, the mean rank for the ith sample, R_i, is given by:

$$R_i = \frac{1}{K_i} \sum_{j=1}^{K_i} r_j \qquad [3.35]$$

The values of the R_i's for our example groups are given in Table 3.6. Once the R_i values are calculated for each group, we calculate our test statistic H as:

$$H = \left[\frac{12}{(N^2+N)} \sum_{i=1}^{M} K_i R_i^2 \right] - 3(N+1) \qquad [3.36]$$

The value of H for our example is 22.18, which has an approximate P-value of 0.0046, indicating a statistically significant difference among the days as was the case with the parametric ANOVA.

If there are tied values we also calculate a correction term C by first counting the number of entries E_q in each if the V tied groups. For example if we had 3 tied groups with 3, 2, and 4 members each we would have $E_1 = 3$, $E_2 = 2$, and $E_3 = 4$. We then compute T_q for each tied group as:

$$T_q = E_q^3 - E_q \qquad [3.37]$$

Our correction term, C, is given by:

$$C = 1 - \frac{\sum\limits_{q=1}^{v} T_q}{(N^3 - N)} \qquad [3.38]$$

Our tie corrected H value, H_C, is given by:

$$H_C = H/C \qquad [3.39]$$

H_C (or simply H in the case of no ties) is compared to a chi-squared statistic with M-1 degrees of freedom.

Multiple Comparisons: Which Means are Different?

When a parametric or nonparametric ANOVA rejects the null hypothesis that all means are the same, the question "Which means are different?" almost inevitably arises. There is a very broad literature on multiple comparisons (e.g., Miller, 1981), but we will focus on a single approach usually called the Bonferroni method for multiple comparisons.

The problem it addresses is that when one has many comparisons the probability P of one or more "significant differences" is given by $1 - (1 - \alpha)^Q$, where Q is the number of multiple comparisons. In general if we have M groups among which pair wise comparison are to be made, the total number of comparisons, Q, is the number of combinations of M things taken 2 at a time.

$$Q = \frac{M(M-1)}{2} \qquad [3.40]$$

Thus, if we did all possible pair-wise comparisons for days in our residual pesticide example, we would have $(9 \bullet 8/2) = 36$ possible pair-wise comparisons.

The probability of one or more chance significant results when $\alpha = 0.05$ for each comparison is $1 - 0.95^{36}$ or 0.84. That is we are quite likely to see one or more significant differences, even if no real differences exist. The cure for this problem is to select a new Bonferroni significance level, α_B, such that:

$$(1 - \alpha_B)^M = (1 - \alpha)$$

or $\qquad\qquad\qquad\qquad\qquad\qquad\qquad\qquad\qquad\qquad\qquad$ [3.41]

$$\alpha_B = 1 - (1 - \alpha)^{1/M}$$

Thus for our example, if our desired overall α is 0.05, $\alpha_B = 1 - 0.95^{1/36} = 0.001423$. We therefore would only consider significant pair-wise differences to be those with a P value of 0.001423 or less.

The actual tests used in this pairwise comparison are pairwise t-tests with S_D taken as the square root of the result of Equation [3.11] with S_p^2 equal to the within-group mean square from the ANOVA. For the alternative Kruskal-Wallis analysis, the test is simply the two-sample U statistic for the two groups. We note that sometimes people look at Equation [3.41] and believe they can "save" on the number of comparisons and thus more easily demonstrate statistical significance by focusing on "interesting" (read large) mean difference comparisons.

The problem is if we pick the differences because they are big, we are biasing our test. That is, we will see more significant differences than really exist. There are correct procedures like the Student-Newman-Keuls test that test largest differences first (Sokol and Rohlf, 1995), and others like Dunnett's test (Zar, 1996) that compare several treatment groups against a common control. For the sorts of statistical prospecting expeditions that characterize many environmental quality problems, the Bonferroni method is a simple and fairly robust tool.

We also note that many reports on environmental quality investigations are filled with large numbers of tables that contain even larger numbers of hypothesis tests. In general the significant differences reported are ordinary pairwise comparisons. After reading this section we hope that our audience has a better appreciation of the fact that such reported significant differences are, at best, to be considered indications of possible differences, rather than results that have true statistical significance.

References

Beyer, W. H. (ed.), 1966, *Handbook of Tables for Probability and Statistics*, CRC Press, Cleveland, OH.

Brown, D. R., Michels, K. M., and Winer, B. J., 1991, *Statistical Principles in Experimental Design*, McGraw Hill, New York.

Conover, W. J., 1998, *Practical Nonparametric Statistics*. John Wiley, New York.

Freedman, D. A., Pisani, R., and Purves, R., 1997, *Statistics, 3rd ed.*, W. W. Norton & Company, New York.

Hogg, R. V. and Craig, A. T., 1995. *An Introduction to Mathematical Statistics, 5th Edition*, Prentice Hall, Englewood Cliffs, NJ.

Kruskal, W. H. and Wallis, W. A., 1952, "Use of Ranks in One-Criterion Analysis of Variance," *Journal of the American Statistical Association*, 47: 583–421.

Lehmann, E. L, 1998, *Nonparametrics: Statistical Methods Based on Ranks*, Prentice Hall, Englewood Cliffs, NJ.

Miller, R. G., 1981, *Simultaneous Statistical Inference*, Springer-Verlag, New York.

Moore, D. S. and McCabe, G. P., *Introduction to the Practice of Statistics, 2nd ed.*, W. H. Freeman and Co., New York.

Sokol, R. R. and Rohlf, F. J., 1995, *Biometry*, W. H. Freeman, New York.

Stigler, S. M., 1986, *The History of Statistics: The Measurement of Uncertainty before 1900*, The Belknap Press of Harvard University Press, Cambridge, MA.

USEPA, 1989, *Methods for Evaluating the Attainment of Cleanup Standards. Volume 1: Soils and Solid Media*, Washington, D.C., EPA 230/02-89-042.

USEPA, 1994a, *Statistical Methods for Evaluating the Attainment of Cleanup Standards. Volume 3: Reference-Based Standards for Soils and Solid Media*, Washington, D.C., EPA 230-R-94-004.

USEPA, 1994b, *Guidance for the Data Quality Objectives Process*, EPA QA/G-4.

USEPA, 1994c, *Data Quality Objectives Decision Error Feasibility Trials (DQO/DEFT), User's Guide*, Version 4, EPA QA/G-4D.

U.S. Nuclear Regulatory Commission, 1995, *A Nonparametric Statistical Methodology for the Design and Analysis of Final Status Decommissioning Surveys*, NUREG-1505.

Zar, J. H., 1996, *Biostatistical Analysis*, Prentice Hall, Englewood Cliffs, NJ.

CHAPTER 4

Correlation and Regression

"Regression is not easy, nor is it fool-proof. Consider how many fools it has so far caught. Yet it is one of the most powerful tools we have — almost certainly, when wisely used, the single most powerful tool in observational studies.

Thus we should not be surprised that:

(1) Cochran said 30 years ago, "Regression is the worst taught part of statistics."
(2) He was right then.
(3) He is still right today.
(4) We all have a deep obligation to clear up each of our own thinking patterns about regression."

(Tukey, 1976)

Tukey's comments on the paper entitled "Does Air Pollution Cause Mortality?" by Lave and Seskin (1976) continues with "difficulties with causal certainty CANNOT be allowed to keep us from making lots of fits, and from seeking lots of alternative explanations of what they might mean."

"For the most environmental [problems] health questions, the best data we will ever get is going to be unplanned, unrandomized, observational data. Perfect, thoroughly experimental data would make our task easier, but only an eternal, monolithic, infinitely cruel tyranny could obtain such data."

"We must learn to do the best we can with the sort of data we have ...."

It is not our intent to provide a full treatise on regression techniques. However, we do highlight the basic assumptions required for the appropriate application of linear least squares and point out some of the more common foibles frequently appearing in environmental analyses. The examples employed are "real world" problems from the authors' consulting experience. The highlighted cautions and limitations are also as a result of problems with regression analyses found in the real world.

Correlation and Regression: Association between Pairs of Variables

In Chapter 2, we introduced the idea of the variance (Equation [2.10]) of a variable x. If we have two variables, x and y, for each of N samples, we can calculate the sample covariance, C_{xy}, as

$$C_{xy} = \frac{\sum_{i=1}^{N} (x_i - \bar{x})(y_i - \bar{y})}{(N-1)} \qquad [4.1]$$

This is a measure of the linear association between the two variables. If the two variables are entirely independent, $C_{xy} = 0$. The maximum and minimum values for C_{xy} are a function of the variability of x and y. If we "standardize" C_{xy} by dividing it by the product of the sample standard deviations (Equation [2.12]) we get the Pearson product-moment correlation coefficient, r:

$$r = C_{xy}/(S_x S_y) \qquad [4.2]$$

The correlation coefficient ranges from −1, which indicates perfect negative linear association, to +1, which indicates perfect positive linear association. The correlation can be used to test the linear association between two variables when the two variables have a bivariate normal distribution (e.g., both x and y are normally distributed). Table 4.1 shows critical values of r for samples ranging from 3 to 50.

For sample sizes greater than 50, we can calculate the Z transformation of r as:

$$Z = \frac{1}{2}\ln\left(\frac{1+r}{1-r}\right) \qquad [4.3]$$

For large samples, Z has an approximate standard deviation of $1/(N-3)^{1/2}$. The expectation of Z under H_0, $\rho = 0$, where ρ is the "true" value of the correlation coefficient. Thus, Z_S, given by:

$$Z_S = Z\sqrt{N-3} \qquad [4.4]$$

is distributed as a standard normal variate, and [4.4] can be used to calculate probability levels associated with a given correlation coefficient.

Spearman's Coefficient of Rank Correlation

As noted above, the Pearson correlation coefficient measures linear association, and the hypothesis test depends on the assumption that both x and y are normally distributed. Sometimes, as shown in Panel A of Figure 4.1, associations are not linear. The Pearson correlation coefficient for Panel A is about 0.79 but the association is not linear.

One alternative is to replace the rank x and y variables from smallest to largest (separately for x and y; for tied values each value in the tied set is assigned the average rank for the tied set and calculate the correlation using the ranks rather than the actual data values. This procedure is called Spearman's coefficient of rank correlation. Approximate critical values for the Spearman rank correlation coefficient are the same as those for the Pearson coefficient and are also given in Table 4.1, for sample sizes of 50 and less. For samples greater than 50, the Z transformation shown in Equations [4.3] and [4.4] can be used to calculate probability levels.

Table 4.1
Critical Values for Pearson and Spearman Correlation Coefficients

No. Pairs	$\alpha = 0.01$	$\alpha = 0.05$	No. Pairs	$\alpha = 0.01$	$\alpha = 0.05$
3	-	0.997	16	0.623	0.497
4	0.990	0.950	17	0.606	0.482
5	0.959	0.878	18	0.59	0.468
6	0.917	0.811	19	0.575	0.456
7	0.875	0.754	20	0.561	0.444
8	0.834	0.707	21	0.549	0.433
9	0.798	0.666	22	0.537	0.423
10	0.765	0.632	25	0.505	0.396
11	0.735	0.602	30	0.463	0.361
12	0.708	0.576	35	0.43	0.334
13	0.684	0.553	40	0.403	0.312
14	0.661	0.532	45	0.38	0.294
15	0.641	0.514	50	0.361	0.279

Critical values obtained using the relationship $t = (N - 2)^{\frac{1}{2}} r/(1 + r^2)^{\frac{1}{2}}$, where t comes from the "t"-distribution. The convention is employed by SAS®.

Bimodal and Multimodal Data: A Cautionary Note

Panel C in Figure 4.1 shows a set of data that consist of two "clumps." The Pearson correlation coefficient for these data is about 0.99 (e.g., nearly perfect) while the Spearman correlation coefficient is about 0.76. In contrast, the Pearson and Spearman correlations for the upper "clump" are 0.016 and 0.018, and for the lower clump are −0.17 and 0.018, respectively. Thus these data display substantial or no association between x and y depending on whether one considers them as one or two samples.

Unfortunately, data like these arise in many environmental investigations. One may have samples upstream of a facility that show little contamination and other samples downstream of a facility that are heavily contaminated. Obviously one would not use conventional tests of significance to evaluate these data (for the Pearson correlation the data are clearly not bivariate normal), but exactly what one should do with such data is problematic. We can recommend that one always plot bivariate data to get a graphical look at associations. We also suggest that if one has a substantial number of data points, one can look at subsets of the data to see if the parts tell the same story as the whole.

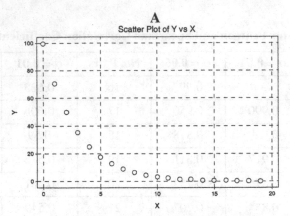

Figure 4.1A   Three Forms of Association
(A is Exponential)

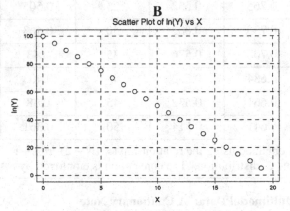

Figure 4.1B   Three Forms of Association
(B is Linear)

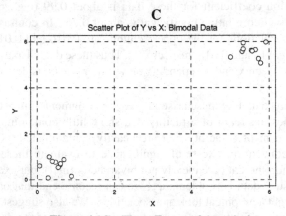

Figure 4.1C   Three Forms of Association
(C is Bimodal)

For the two clumps example, one might wish to examine each clump separately. If there is substantial agreement between the parts analyses and the whole analysis, one's confidence on the overall analysis is increased. On the other hand, if the result looks like our example, one's interpretation should be exceedingly cautious.

Linear Regression

Often we are interested in more than simple association, and want to develop a linear equation for predicting y from x. That is we would like an equation of the form:

$$\hat{y}_i = \hat{\beta}_0 + \hat{\beta}_1 x_i \qquad [4.5]$$

where $\hat{y}_i$ is the predicted value of the mean of y for a given x,

$$\mu_{y|x} = \beta_0 + \beta_1 x$$

and β_0 and β_1 are the intercept and slope of the regression equation. To obtain an estimate of β_1, we can use the relationship:

$$\hat{B}_1 = C_{xy} / S_x^2 \qquad [4.6]$$

The intercept is estimated as:

$$\hat{\beta}_0 = \bar{y} - \hat{\beta}_1 \bar{x} \qquad [4.7]$$

We will consider in the following examples several potential uses for linear regression and while considering these uses, we will develop a general discussion of important points concerning regression. First, we need a brief reminder of the often ignored assumptions permitting the linear "least squares" estimators, $\hat{\beta}_0$ and $\hat{\beta}_1$, to be the minimum variance linear unbiased estimators of β_0 and β_1, and, consequently $\hat{y}_I$, to be the minimum variance linear unbiased estimator of $\mu_{y|x}$. These assumptions are:

- The values of x are known without error.
- For each value of x, y is independently distributed with $\mu_{y|x} = \beta_0 + \beta_1 x$ and variance $\sigma_{y|x}^2$.
- For each x the variance of y given x is the same; that is $\sigma_{y|x}^2 = \sigma^2$ for all x.

Calculation of Residue Decline Curves

One major question that arises on the course of environmental quality investigations is residue decline. That is, we might have toxic material spilled at an industrial site, PCBs, and dioxins in aquatic sediments, or pesticides applied to crops. In each case the question is the same: "Given that I have toxic material in the environment, how long will it take it to go away?" To answer this question we perform a linear regression of chemical concentrations, in samples taken at different times postdeposition, against the time that these samples were collected. We will consider three potential models for residue decline.

Exponential:

$$C_t = C_0 e^{-\beta_1 t}$$

or

$$\ln(C_t) = \beta_0 - \beta_1 t \qquad [4.8]$$

Here C_t is the concentration of chemical at time t, which is equivalent to $\hat{y}_1$, β_0 is an estimate of $\ln(C_0)$, the log of the concentration at time zero, derived from the regression model, and β_1 is the decline coefficient that relates change in concentration to change in time.

Log-log:

$$C_t = C_0(1 + t)^{-\beta_1}$$

or

$$\ln(C_t) = \beta_0 - \beta_1 \ln(1 + t) \qquad [4.9]$$

Generalized:

$$C_t = C_0(1 + \Phi t)^{-\beta_1}$$

or

$$\ln(C_t) = \beta_0 - \beta_1 \ln(1 + \Phi t) \qquad [4.10]$$

In each case we are evaluating the natural log of concentration against a function of time. In Equations [4.7] and [4.8], the relationship between $\ln(C_t)$ and either time or a transformation of time is the simple linear model presented in Equation [4.5]. The relationship in [4.10] is inherently nonlinear because we are estimating an additional parameter, Φ. However, the nonlinear solution to [4.10] can be found by using linear regression for multiple values of Φ and picking the Φ value that gives the best fit.

Exponential Decline Curves and the Anatomy of Regression

The process described by [4.8] is often referred to as exponential decay, and is the most commonly encountered residue decline model. Example 4.1 shows a residue decline analysis for an exponential decline curve. The data are in the first panel. The analysis is in the second. The important feature here is the regression analysis of variance. The residual or error sum of squares, SS_{RES}, is given by:

$$SS_{RES} = \sum_{i=1}^{N} (y_i - \hat{y})^2 \qquad [4.11]$$

Example 4.1 A Regression Analysis of Exponential Residue Decline

Panel 1. The Data

Time (t)	Residue(C_t)	ln(Residue)
0	157	5.05624581
2	173	5.15329159
4	170	5.13579844
8	116	4.75359019
11	103	4.63472899
15	129	4.8598124
22	74	4.30406509
29	34	3.52636052
36	39	3.66356165
43	35	3.55534806
50	29	3.36729583
57	29	3.36729583
64	17	2.83321334

Panel 2. The Regression Analysis

```
Linear Regression of ln(residue) versus time

Predictor              Standard
Variable      β        Error of β (Sβ)    Student's t p-value

ln(C₀)     5.10110     0.09906               51.49     0.0000
time      -0.03549     0.00294              -12.07     0.0000

R-SQUARED = 0.9298

ANOVA Table for Regression

SOURCE            DF        SS        MS       F      P

REGRESSION         1     7.30763   7.30763 145.62 0.0000
RESIDUAL          11     0.55201   0.05018
TOTAL             12     7.85964
```

Panel 3. The Regression Plot

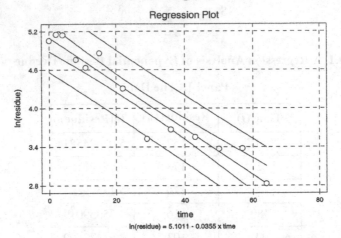

Regression Plot

ln(residue) = 5.1011 - 0.0355 x time

Panel 4. Calculation of Prediction Bounds Time = 40

a. $\bar{y} = 4.17$ b. $y' = 4.408$ c. $\beta = \beta_1 = -0.03549$

d. $T = t_{11,0.95} = 1.796$ e. $S_\beta = 0.00294$ f. $Q = \beta_1^2 - T^2 S_\beta^2$

g. $\Sigma(x - \bar{x})^2 = 5800.32$ h. $E = (y' - \bar{y})^2 / \Sigma(x - \bar{x})^2$ i. $G = (N + 1)/N$

j. $V = \bar{x} + \{\beta_1(y' - \bar{y})/Q$

       $= 26.231 - 0.03549 \cdot (4.408 - 4.17)/0.00123166 = 19.3731$

k. $x' = (y' - \beta_0)/\beta_1 = (4.408 - 5.10110)/-0.03549 = 19.53$

l. $D = T/Q\{S_{y \cdot x}^2 (E + QG)\}^{1/2}$

       $= (1.796/0.00123166) \cdot (0.05018 \cdot (4.103 \times 10^{-5} + 0.00123166 \cdot 1.07692))^{1/2}$

       $= 12.0794$

m. $L1 = V - D = 19.3731 - 12.0794 = 7.2937$

n. $L2 = V + D = 19.3731 + 12.0794 = 31.4525$

Panel 5. Calculation of the Half Life and a Two-Sided 90% Confidence Interval

$S_{y \cdot x}^2$ Residual mean square = 0.05018 $S_{y \cdot x} = 0.224$

$S(\hat{y})$ (standard error of $\hat{y}_i$) $= S_{y \cdot x}[1 + (1/N) + \{(x_i - \bar{x})^2 / \Sigma(x - \bar{x})^2\}]^{1/2}$

$S(\hat{y}) = 0.224 [1 + 1/13 + \{(40 - 26.231)^2/5836.32\}]^{1/2} = 0.2359$

95% UB $= \hat{y}_i + t_{N-2(0.975)} S(\hat{y}) = 3.6813 + 2.201 - 0.2359 = 4.20$

95% LB $= \hat{y}_i + t_{N-2(0.975)} S(\hat{y}) = 3.6813 - 2.201 - 0.2359 = 3.16$

In original units (LB, Mean, UB): 23.57, 39.70, 66.69

The total sum of squares, SS_{TOT}, is given by:

$$SS_{TOT} = \sum_{i=1}^{N} (y_1 - \bar{y})^2 \qquad [4.12]$$

The regression sum of squares, SS_{REG}, is found by subtraction:

$$SS_{REG} = SS_{TOT} - SS_{RES} \qquad [4.13]$$

The ratio of SS_{REG}/SS_{TOT} is referred to as the R^2 value or the explained variation. It is equal to the square of the Pearson correlation coefficient between x and y. This is the quantity that is most often used to determine how "good" a regression analysis is. If one is interested in precise prediction, one is looking for R^2 values of 0.9 or so. However, one can have residue decline curves with much lower R^2 values (0.3 or so) which, though essentially useless for prediction, still demonstrate that residues are in fact declining.

In any single variable regression, the degrees of freedom for regression is always 1, and the residual and total degrees of freedom are always $N-2$ and $N-1$, respectively. Once we have our sums of squares and degrees of freedom we can construct mean squares and an F-test for our regression. Note that the regression F tests a null hypothesis (H_0) of $\beta_1 = 0$ versus an alternative hypothesis (H_1) of $\beta_1 \neq 0$. For things like pesticide residue studies, this is not a very interesting test because we know residues are declining with time. However, for other situations like PCBs in fish populations or river sediments, it is often a question whether or not residues are actually declining. Here we have a one-sided test where H_0 is $\beta_1 \geq 0$ versus an H_1 of $\beta_1 < 0$. Note also that most regression programs will report standard errors (s_β) for the β's. One can use the ratio β/s_β to perform a t-test. The ratio is compared to a t statistic with $N-2$ degrees of freedom.

Prediction is an important problem. A given $\hat{y}$ can be calculated for any value of x. A confidence interval for a single y observation for a given $\hat{y}$ value is shown in Panel 4 of Example 4.1. This is called the prediction interval. A confidence interval for $\hat{y}$ is $C(y)$ given by:

$$C(\hat{y}) = \hat{y}_j + t_{(N-2,1-\alpha/2)} \, S_{yx} \left[\left(\frac{1}{N}\right) + \frac{(x_j - \bar{x})^2}{\sum_{i=1}^{N}(x_i - \bar{x})^2} \right]^{1/2} \qquad [4.14]$$

The difference between these two intervals is that the prediction interval is for a new y observation at a particular x, while the confidence interval is for $\mu_{y|x}$ itself.

One important issue is inverse prediction. That is, in terms of residue decline we might want to estimate the time (our x variable) environmental residues (our y variable) to reach a given level y′. To do this we "invert" Equation 4.5; that is:

$$y' = \beta_0 + \beta_1 x', \text{ or, } x' = (y' - \beta_0) / \beta_1 \qquad [4.15]$$

For an exponential residue decline problem, calculation of the "half-life" (the time that it takes for residues to reach 1/2 their initial value) is often an important issue. If we look at Equation [4.15], it is clear that the half-life (H) is given by:

$$H = \ln(0.5) / \beta_1 \qquad [4.16]$$

because y′ is the log of 1/2 the initial concentration and β_0 is the log of the initial concentration.

For inverse prediction problems, we often want to calculate confidence intervals for the predicted x′ value. That is, if we have, for example, calculated a half-life estimate, we might want to set a 95% upper bound on the estimate, because this value would constitute a "conservative" estimate of the half-life. Calculation of a 90% confidence interval for the half-life (the upper end of which corresponds to a 95% one-sided upper bound) is illustrated in Panel 4 of Example 4.1. This is a quite complex calculation.

If one is using a computer program that calculates prediction intervals, one can also calculate approximate bounds by finding L1 as the x value whose 90% (generally, $1 - \alpha$; the width of the desired two-sided interval) two-sided lower prediction bound equals y′ and L2 as the x value whose 90% two-sided upper prediction bound equals y′. To find the required x values one makes several guesses for L# (here # is 1 or 2) and finds two that have $L\#_1$ and $L\#_2$ values for the required prediction bounds that bracket y′. One then calculates the prediction bound for a value of L# intermediate between $L\#_1$ and $L\#_2$. Then one determines if y′ is between $L\#_1$ and the bound calculated from the new L# or between the new L# and $L\#_2$.

In the first case L# becomes our new $L\#_2$ and in the second L# becomes our new $L\#_1$. We then repeat the process. In this way we confine the possible value of the desired L value to a narrower and narrower interval. We stop when our L# value gives a y value for the relevant prediction bound that is acceptably close to y′. This may sound cumbersome, but we find that a few guesses will usually get us quite close to y′ and thus L1 or L2. Moreover, if the software automatically calculates prediction intervals (most statistical packages do), its quite a bit easier than setting up the usual calculation (which many statistical packages do not do) in a spreadsheet. For our problem these approximate bounds are 7.44 and 31.31, which agree pretty well with the more rigorous bounds calculated in Panel 4 of Example 4.1.

Other Decline Curves

In Equations [4.9] and [4.10] we presented two other curves that can be used to describe residue decline. The log-log model is useful for fitting data where there are several compartments that have exponential processes with different half-lives. For example, pesticides on foliage might have a surface compartment from which material dissipates rapidly, and an absorbed compartment from which material dissipates relatively slowly.

All of the calculations that we did for the exponential curve work the same way for the log-log curve. However, we can calculate a half-life for an exponential curve and can say that, regardless where we are on the curve, the concentration after one half-life is one-half the initial concentration. That is, if the half-life is three days, then concentration will drop by a factor of 2 between day 0 and day 3, between day 1 and day 4, or day 7 and day 10. For the log-log curve we can calculate a time for one-half of the initial concentration to dissipate, but the time to go from 1/2 the initial concentration to 1/4 the initial concentration will be much longer (which is why one fits a log-log as opposed to a simple exponential model in the first place).

The nonlinear model shown in [4.10] (Gustafson and Holden, 1990) is more complex. When we fit a simple least-squares regression we will always get a solution, but for a nonlinear model there is no such guarantee. The model can "fail to converge," which means that the computer searches for a model solution but does not find one. The model is also more complex because it involves three parameters, β_0, β_1, and Φ. In practice, having estimated Φ we can treat it as a transformation of time and use the methods presented here to calculate things like prediction intervals and half-times. However, the resulting intervals will be a bit too narrow because they do not take the uncertainty in the Φ estimate into account.

Another problem that can arise from nonlinear modeling is that we do not have the simple definition of R^2 implied by Equation [4.13]. However, any regression model can calculate an estimate $(\hat{y})$ for each observed y value, and the square of the Pearson product-moment correlation coefficient, r, between y_i and $\hat{y}_i$, which is exactly equivalent to R^2 for least-squares regression (hence the name R^2) can provide an estimate comparable to R^2 for any regression model.

We include the nonlinear model because we have found it useful for describing data that both exponential and simple log-log models fail to fit and because nonlinear models are often encountered in models of residue (especially soil residue) decline.

Regression Diagnostics

In the course of fitting a model we want to determine if it is a "good" model and/or if any points have undue influence on the curve. We have already suggested that we would like models to be predictive in the sense that they have a high R^2, but we would also like to identify any anomalous features of our data that the decline regression model fails to fit. Figure 4.2 shows three plots that can be useful in this endeavor.

Plot A is a simple scatter plot of residue versus time. It suggests that an exponential curve might be a good description of these data. The two residual plots show the residuals versus their associated $\hat{y}_i$ values. In Plot B we deliberately fit a linear model, which Plot A told us would be wrong. This is a plot of "standardized" residuals $(y_i - \hat{y}_i)$ versus fitted values $\hat{y}_i$ for a regression of residue on time. The standardized residuals are found by subtracting mean dividing by the standard deviation of the residuals. The definite "V" shape in the plot shows that there are systematic errors on the fit of our curve.

Plot C is the same plot as B but for the regression of ln(residue) on time. Plot A shows rapid decline at first followed by slower decline. Plot C, which shows residuals versus their associated $\hat{y}_i$ values, has a much more random appearance, but suggests one possible outlier. If we stop and consider Panel 3 of Example 4.1, we see that the regression plot has one point outside the prediction interval for the regression line, which further suggests an outlier.

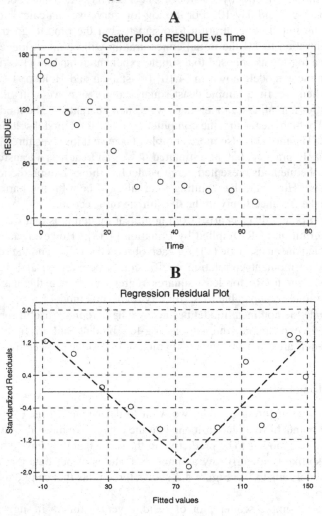

Figure 4.2 Some Useful Regression Diagnostic Plots

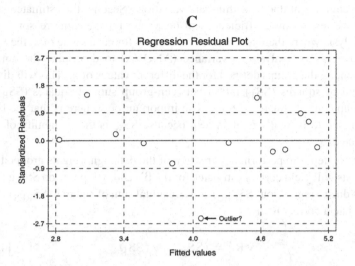

Figure 4.2   Some Useful Regression Diagnostic Plots (Cont'd)

The question that arises is: "Did this outlier influence our regression model?" There is substantial literature in identifying problems in regression models (e.g., Belsley, Kuh, and Welsch, 1980) but the simplest approach is to omit a suspect observation from the calculation, and see if the model changes very much. Try doing this with Example 4.1. You will see that while the point with the large residual is not fit very well, omitting it does not change our model much.

One particularly difficult situation is shown in Figure 4.1C. Here, the model will have a good R^2 and omitting any single point will have little effect on the overall model fit. However, the fact remains that we have effectively two data points, and as noted earlier, any line will do a good job of connecting two points. Here our best defense is probably the simple scatter plot. If you see a data set where there are, in essence, a number of tight clusters, one could consider the data to be grouped (see below) or try fitting separate models within groups to see if they give similar answers. The point here is that one cannot be totally mechanical in selecting regression models; there is both art and science in developing good description of the data.

Grouped Data: More Than One y for Each x

Sometimes we will have many observations of environmental residues taken at essentially the same time. For example, we might monitor PCB levels in fish in a river every three months. On each sample date we may collect many fish, but the date is the same for each fish at a given monitoring period. A pesticide residue example is shown in Example 4.2.

If one simply ignores the grouped nature of the data one will get an analysis with a number of errors. First, the estimated R^2 will be not be correct because we are looking at the regression sum of squares divided by the total sum of squares, which includes a component due to within-date variation. Second, the estimated standard errors for the regression coefficients will be wrong for the same reason. To do a correct analysis where there are several values of y for each value of x, the first step is to do a one-way analysis of variance (ANOVA) to determine the amount of variation among the groups defined for the different values of x. This will divide the overall sum of squares (SS_T) into a between-group sum of squares (SS_B) and a within-group sum of squares (SS_W). The important point here is that the best any regression can do is totally explain SS_B because SS_W is the variability of y's at a single value of x.

The next step is to perform a regression of the data, ignoring its grouped nature. This analysis will yield correct estimates for the β's and will partition SS_T into a sum of squares due to regression (SS_{REG}) and a residual sum of squares (SS_{RES}). We can now calculate a correct R^2 as:

$$R^2 = (SS_{REG}) / (SS_B) \qquad [4.17]$$

Example 4.2 Regression Analysis for Grouped Data

Panel 1. The Data

Time	Residue	ln(Residue)	Time	Residue	ln(Residue)
0	3252	8.08703	17	548	6.30628
0	3746	8.22844	17	762	6.63595
0	3209	8.07371	17	2252	7.71957
1	3774	8.23589	28	1842	7.51861
1	3764	8.23323	28	949	6.85541
1	3211	8.07434	28	860	6.75693
2	3764	8.23324	35	860	6.75693
2	5021	8.52138	35	1252	7.13249
2	5727	8.65295	35	456	6.12249
5	3764	8.23324	42	811	6.69827
5	2954	7.99092	42	858	6.75460
5	2250	7.71869	42	990	6.89770
7	2474	7.81359	49	456	6.12249
7	3211	8.07434	49	964	6.87109
7	3764	8.23324	49	628	6.44254

Panel 2. The Regression

Linear regression of ln(RESIDUE) versus TIME: Grouped data

PREDICTOR VARIABLE	β	STD ERROR (β	STUDENT'S T	P
CONSTANT	8.17448	0.10816	75.57	0.0000
TIME	-0.03806	0.00423	-9.00	0.0000

R-SQUARED = 0.7431

ANOVA Table for Regression

SOURCE	DF	SS	MS	F	P
REGRESSION	1	13.3967	13.3967	81.01	0.0000
RESIDUAL	28	4.63049	0.16537		
TOTAL	29	18.0272			

Panel 3. An ANOVA of the Same Data

One-way ANOVA for ln(RESIDUE) by time

SOURCE	DF	SS	MS	F	P
BETWEEN	9	15.4197	1.71330	13.14	0.0000
WITHIN	20	2.60750	0.13038		
TOTAL	29	18.0272			

Panel 4. A Corrected Regression ANOVA, with Corrected R^2

Corrected regression ANOVA

SOURCE	DF	SS	MS	F	P
REGRESSION	1	13.3967	13.3967	52.97	0.0000
LACK OF FIT	8	2.0230	0.2529	1.94	0.1096
WITHIN	20	2.6075	0.1304		
TOTAL	29	18.0272			

R^2 = REGRESSION SS/BETWEEN SS = 0.87

We can also find a lack-of-fit sum of squares (SS_{LOF}) as:

$$SS_{LOF} = SS_B - SS_{REG} \qquad [4.18]$$

We can now assemble the corrected ANOVA table shown in Panel 4 of Example 4.2 because we can also find our degrees of freedom by subtraction. That is, SS_{REG} has one degree of freedom and SSB has $K - 1$ degrees of freedom (K is the number of groups), so SS_{LOF} has $K - 2$ degrees of freedom. Once we have the correct sums of squares and degrees of freedom we can calculate mean squares and F tests. Two F tests are of interest. The first is the regression F (F_{REG}) given by:

$$F_{REG} = MS_{REG}/MS_{LOF} \qquad [4.19]$$

The second is a lack of fit F (F_{LOF}), given by:

$$F_{LOF} = MS_{LOF}/MS_W$$

If we consider the analysis in Example 4.1, we began with an R^2 of about 0.74, and after we did the correct analysis found that the correct R^2 is 0.87. Moreover the F_{LOF} says that there is no significant lack of fit in our model. That is, given the variability of the individual observations we have done as well as we could reasonably expect to. We note that this is not an extreme example. We have seen data for PCB levels in fish where the initial R^2 was around 0.25 and the regression was not significant, but when grouping was considered, the correct R^2 was about 0.6 and the regression was clearly significant. Moreover the F_{LOF} showed that given the high variability of individual fish, our model was quite good. Properly handling grouped data in regression is important.

One point we did not address is calculation of standard errors and confidence intervals for the β's. If, as in our example, we have the same number of y observations for each x, we can simply take the mean of the y's at each x and proceed as though we had a single y observations for each x. This will give the correct estimates for R^2 (try taking the mean ln(Residue) value for each time in Example 4.1 and doing a simple linear regression) and correct standard errors for the β's. The only thing we lose is the lack of fit hypothesis test. For different numbers of y observations for each x, the situation is a bit more complex. Those needing information about this can consult one of several references given at the end of this chapter (e.g., Draper and Smith, 1998; Sokol and Rolhf, 1995; Rawlings, Pantula, and Dickey, 1998).

Another Use of Regression: Log-Log Models for Assessing Chemical Associations

When assessing exposure to a mix of hazardous chemicals, the task may be considerably simplified if measurements of a single chemical can be taken as a surrogate or indicator for another chemical in the mixture. If we can show that the concentration of chemical A is some constant fraction, F, of chemical B, we can measure the concentration of B, C_B, and infer the concentration of A, C_A, as:

$$C_A = F \bullet C_B \qquad [4.20]$$

One can use the actual measurements of chemicals A and B to determine whether a relationship such as that shown in [4.20], in fact, exists.

Typically, chemicals in the environmental are present across a wide range of concentrations because of factors such as varying source strength, concentration and dilution in environmental media, and chemical degradation. Often the interaction of these factors acts to produce concentrations that follow a log-normal distribution. The approach discussed here assumes that the concentrations of chemicals A and B follow log-normal distributions.

If the concentration of a chemical follows a log-normal distribution, the log of the concentration will follow a normal distribution. For two chemicals, we expect a bivariate log-normal distribution, which would translate to a bivariate normal distribution for the log-transformed concentrations. If we translate [4.20] to logarithmic units we obtain:

$$\ln(C_A) = \ln(F) + \ln(C_B) \qquad [4.21]$$

This the regression equation of the logarithm of C_A on the logarithm of C_B. That is, when $\ln(C_A)$ is the dependent variable and $\ln(C_B)$ is the independent variable, the regression equation is:

$$\ln(C_A) = \beta_0 + \beta_1 \ln(C_B) \qquad [4.22]$$

If we let $\ln(F) = \beta_0$, (i.e., $F = e^{\beta_0}$) and back-transform [4.22] to original units by taking exponentials (e.g., e^x where X is any regression term of interest), we obtain:

$$C_A = F C_B^{\beta_1} \qquad [4.23]$$

This [4.23] is the same as [4.20] except for the β exponential term on C_B, and [4.23] would be identical to [4.20] for the case $\beta_1 = 1$.

Thus, one can simply regress the log-transformed concentrations of one chemical on the log-transformed concentration of the other chemical (assuming that the pairs of concentrations are from the same physical sample). One can then use the results of this calculation to evaluate the utility of chemical B as an indicator for chemical A by statistically testing whether $\beta_1 = 1$. This is easily done with most statistical packages because they report the standard error of β_1 and one can thus calculate a confidence interval for β_1 as in our earlier examples. If this interval includes 1, it follows that C_A is a constant fraction of C_B and this fraction is given by F.

For a formal test of whether Equation [4.21] actually describes the relationship between chemical A and chemical B, one proceeds as follows:

1. Find the regression coefficient (β) for Log (chemical B) regressed on Log (chemical A) together with the standard error of this coefficient (SE_β). (See the examples in the tables.)

2. Construct a formal hypothesis test of whether β equals one as follows:

$$t = (1 - \beta) / (SE_\beta) \qquad [4.24]$$

3.   Compare t to a t distribution with $N - 2$ (N is the number of paired samples) degrees of freedom.

For significance (i.e., rejecting the hypothesis H_0: $\beta = 1$) at the $p = 0.05$ level on a two-sided test (null hypothesis H_0: $\beta = 1$ versus the alternate hypothesis H_1: $\beta \neq 1$), the absolute value of t must be greater than $t_{(N-2, \, 1-\alpha/2)}$. In the event that we fail to reject H_0 (i.e., we accept that $\beta = 1$), it follows that Equation [4.20] is a reasonable description of the regression of A on B and that chemical B may thus be a reasonable linear indicator for chemical A.

An Example

The example in Table 4.2 is taken from a study of exposure to environmental tobacco smoke in workplaces where smoking occurred (LaKind et al., 1999a, 1999b, 1999c). The example considers the log-log regression of the nicotine concentration in air (in $\mu g/m^3$) on the ultraviolet fluorescing particulate matter concentration in air (UVPM; also in $\mu g/m^3$). Here we see that the t statistic described in [4.24] is only 1.91 ($p = 0.06$). Thus, we cannot formally reject H_0, and might wish to consider UVPM as an indicator for nicotine. This might be desirable because nicotine is somewhat harder to measure than UVPM.

However, in this case, the R^2 of the regression model given in Table 4.2 is only 0.63. That is, regression of Log (nicotine) on Log (UVPM) explains only 63 percent of the variation in the log-transformed nicotine concentration. The general regression equation suggests that, on average, nicotine is a constant proportion of UVPM. This proportion is given by $F = 10^{\alpha} = 10^{-1.044} = 0.090$. (Note that we are using log base 10 here rather than log base e. All of the comments presented here are independent of the logarithmic base chosen.) However, the lack of a relatively high R^2 suggests that for individual observations, the UVPM concentration may or may not be a reliable predictor of the nicotine concentration in air. That is, on average the bias is small, but the difference between an individual nicotine level and the prediction from the regression model may be large.

Table 4.2
Regression Calculations for Evaluating the Utility of Ultraviolet Fluorescing Particulate Matter (UVPM) as an Indicator for Nicotine

Predictor Variables	Coefficient	Standard Error	Student's t	P-value
Constant (α)	−1.044	0.034	−30.8	0.00
Log (UVPM) (β)	0.935	0.034	27.9	0.00
R-squared = 0.63			Cases included: 451	

A Caveat and a Note on Errors in Variables Models

In regression models, it is explicitly assumed that the predictor variable (in this case chemical B) is measured without error. Since measured concentrations are in fact estimates based on the outcome of laboratory procedures, this assumption is not met in this discussion. When the predictor variable is measured with error, the slope estimate (β_1) is biased toward zero. That is, if the predictor chemical is measured with error, the β_1 value in our model will tend to be less than 1. However, for many situations the degree of this bias is not large, and we may, in fact, be able to correct for it. The general problem, usually referred to as the "errors in variables problem," is discussed in Rawlings et al. (1998) and in greater detail in Fuller (1987).

One useful way to look at the issue is to assume that each predictor x_i can be decomposed into its "true value," z_i, and an error component, u_i. The u_i's are assumed to have zero mean and variance σ_u^2. One useful result occurs if we assume that (1) the z_i's are normally distributed with mean 0 and variance σ_z^2, (2) the u_i's are normally distributed with mean 0 and variance σ_u^2, and (3) the z_i's and u_i's are independent. Then:

$$\beta_C = \beta_E \bullet (\sigma_z^2 + \sigma_u^2) / \sigma_z^2 \qquad [4.25]$$

where β_C is the correct estimate of β_1, and β_E is the value estimated from the data. It is clear that if σ_z^2 is large compared to σ_u^2. Then:

$$(\sigma_z^2 + \sigma_u^2) / \sigma_z^2 \approx 1 \quad \text{and} \quad \beta_C \approx \beta_E \qquad [4.26]$$

Moreover, we typically have a fairly good idea of σ_u^2 because this is the logarithmic variance of the error in the analytic technique used to analyze for the chemical being used as the predictor in our regression. Also because we assume z_i and u_i to be uncorrelated, it follows that:

$$\sigma_x^2 = \sigma_z^2 + \sigma_u^2 \qquad [4.27]$$

Thus, we can rewrite [4.25] as:

$$\beta_C = \beta_E \bullet \sigma_x^2 / (\sigma_x^2 - \sigma_u^2) \qquad [4.28]$$

How large might this correction be? Well, for environmental measurements, it is typical that 95 percent of the measurements are within a factor of 10 of the geometric mean, and for laboratory measurements we would hope that 95 percent of the measurements would be within 20 percent of the true value.

For log-normal distributions this would imply that on the environmental side:

$$UB_{env,0.975} = GM \bullet 10 \qquad [4.29]$$

That is, the 97.5 percentile upper percentile of the environmental concentration distribution, $UB_{env, 0.975}$, is given by the geometric mean, GM, times ten. If we rewrite [4.29] in terms of logarithms, we get:

$$Log_{10}(UB_{env,0.975}) = Log_{10}(GM) + Log_{10}(10) \qquad [4.30]$$

Here $Log_{10}(GM)$ is the logarithm of the geometric mean, and, of course, in base 10 is 1 ($Log_{10}(10) = 1$). It is also true that:

$$Log_{10}(UB_{env,0.975}) = Log_{10}(GM) + 1.96 \, \sigma_x \qquad [4.31]$$

Thus, equating [4.30] and [4.31]:

$$\sigma_x = Log_{10}(10) / 1.96 = 0.512 \quad \text{and,}$$

$$\text{thus, } \sigma_x^2 = 0.2603 \qquad [4.32]$$

By similar reasoning, for the error distribution attributable to laboratory analysis:

$$UB_{lab,0.975} = GM \bullet 1.2 \qquad [4.33]$$

This results in:

$$\sigma_u = Log_{10}(1.2) / 1.96 = 0.0404 \quad \text{and} \quad \sigma_u^2 = 0.0016 \qquad [4.34]$$

When we substitute the values from [4.32] and [4.34] into [4.28] we obtain:

$$\beta_C = \beta_E \bullet 1.0062 \qquad [4.35]$$

Thus, if 95 percent of the concentration measurements are within a factor of 10 of the geometric mean and the laboratory measurements are within 20 percent of the true values, then the bias in β_E is less than 1 percent.

The first important point that follows from this discussion is that measurement errors usually result in negligible bias. However, if σ_x^2 is small, which would imply that there is little variability in the chemical concentration data, or σ_u^2 is large, which would imply large measurement errors, β_E may be seriously biased toward zero. The points to remember are that if the measurements have little variability or analytic laboratory variation is large, the approach discussed here will not work well. However, for many cases, σ_x^2 is large and σ_u^2 is small, and the bias in β_E is therefore also small.

Calibrating Field Analytical Techniques

The use of alternate analytical techniques capable of providing results rapidly and on site opens the possibility of great economy for site investigation and remediation. The use of such techniques require site-specific "calibration" against standard reference methods. The derivation of this calibrating relationship often involves addressing the issues discussed above. While the names of the companies in this example are fictitious, the reader is advised that the situation, the data, and the statistical problems discussed are very real.

The W. E. Pack and U. G. Ottem Co. packaged pesticides for the consumer market in the 1940s and early 1950s. As the market declined, the assets of Pack and Ottem were acquired by W. E. Stuck, Inc., and operations at the Pack-Ottem site were terminated. The soil at the idle site was found to be contaminated, principally with DDT, during the 1980s. W. E. Stuck, Inc. entered a consent agreement to clean up this site during the early 1990s.

W. E. Stuck, being a responsible entity, wanted to do the "right thing," but also felt a responsibility to its stock holders to clean up this site for as low a cost as possible. Realizing that sampling and analytical costs would be a major portion of cleanup costs, an analytical method other than Method 8080 (the U.S. EPA standard method) for DDT was sought. Ideally, an alternate method would not only cut the analytical costs but also cut the turnaround time associated with the use of an offsite contract laboratory.

The latter criterion has increased importance in the confirmatory stage of site remediation. Here the cost of the idle "big yellow" equipment (e.g., backhoes, front end loaders, etc.) must also be taken into account. If it could be demonstrated that an alternate analytical method with a turnaround time of minutes provided results equivalent to standard methods with a turnaround of days or weeks, then a more cost effective cleanup may be achieved because decisions about remediation can be made on a "real time" basis.

The chemist-environmental manager at W. E. Stuck realized that the mole fraction of the chloride ion (Cl^-) was near 50 percent for DDT. Therefore, a technique for detection of Cl^- such as the Dexsil® L2000 might well provide for the determination of DDT within 15 minutes of sample collection. The Dexsil® L2000 has been identified as a method for the analysis of polychlorinated biphenyls, PCBs, in soil (USEPA, 1993). The method extracts PCBs from soil and dissociates the PCBs with a sodium reagent, freeing the chloride ions.

In order to verify that the Dexsil® L2000 can effectively be used to analyze for DDT at this site, a "field calibration" is required. This site-specific calibration will establish the relationship between the Cl^- concentration as measured by the Dexsil® L2000 and the concentration of total DDT as measured by the reference Method 8080. This calibration is specific for the soil matrix of the site, as it is not known whether other sources of Cl^- are found in the soils at this site.

A significant first step in this calibration process was to make an assessment of the ability of Method 8080 to characterize DDT in the site soil. This established a "lower bound" on how close one might expect a field analysis result to be to a reference method result. It must be kept in mind that the analyses are made on *different* physical samples taken from *essentially* the same location and will likely differ in concentration. This issue was discussed at length in Chapter 1.

Table 4.3 presents the data describing the variation among Method 8080 analyses of samples taken at essentially the same point. Note that the information supplied by these data comes from analyses done as part of the QAPP. Normally these data are relegated to a QA appendix in the project report. One might question the inclusion of "spiked" samples. Usually, these results are used to confirm analytical percent recovery. However, as we know the magnitude of the spike, it is also appropriate to back this out of the final concentration and treat the result as an analysis of another aliquot of the original sample. Note that the pooled standard deviation is precisely equivalent to the square root of the within-group mean square of the ANOVA by the sample identifiers.

Table 4.3
Method 8080 Measurement Variation

Sample Ident.	Total DDT, mg/kg					Degrees of Freedom	Corrected Sum of Squares of Logs
	Original	Dup	Matrix Spike	Matrix Spike Dup	Geom. Mean		
Phase I Samples							
BH-01	470.10		304.60	261.20	334.42	2	0.1858
BH-02	0.25		0.23	0.37	0.28	2	0.1282
BH-03	0.09	0.08			0.08	1	0.0073
BH-04	13.45	5.55			8.63	1	0.3922
BH-05	0.19	0.07			0.12	1	0.4982
BH-06	0.03	0.03			0.03	1	0.0012
BH-07	0.03		0.19	0.21	0.10	2	2.4805
BH-08	1276.00	1544.00			1403.62	1	0.0182
Phase II Samples							
BH-09	130.50	64.90			92.03	1	0.2440
BH-10	370.90	269.70			316.28	1	0.0508
BH-11	635.60	109.10			263.33	1	1.5529
BH-12	0.12	0.30			0.18	1	0.4437
BH-13	41.40	19.59			28.48	1	0.2799
BH-14	12.90	13.50			13.20	1	0.0010
BH-15	4.93	1.51			2.73	1	0.7008
BH-16	186.00	160.30			172.67	1	0.0111
BH-17	15.40	8.62			11.52	1	0.1684
BH-18	10.20	12.37			11.23	1	0.0186
Total =						21	7.1826
Pooled Standard Deviation, S_x =							0.5848

Figure 4.3 presents the individual analyses against their geometric mean. Note that the scale in both directions is logarithmic and that the variation among individual analyses appears to be rather constant over the range. This suggests that the logarithmic transformation of the total DDT data is appropriate. The dashed lines define the 95% prediction interval (Hahn, 1970a, 1970b) throughout the observed range of the data. The upper and lower limits, U_i and L_i, are found for each log geometric mean, $\bar{x}_I$, describing the ith group of repeated measurements. These limits are given by:

$$\left.\begin{array}{c} U_i \\ L_i \end{array}\right\} = \bar{x}_i \pm S_x t_{(N_i - 1, 1 - \alpha/2)} \sqrt{1 + \frac{1}{N_i}} \qquad [4.36]$$

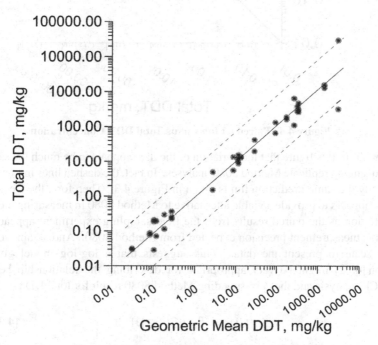

Figure 4.3　Method 8080 Measurement Variation

In order to facilitate the demonstration that the Dexsil Cl⁻ analysis is a surrogate for Method 8080 total DDT analysis, a sampling experiment was conducted. This experiment involved the collection of 49 pairs of samples at the site. The constraints on the sampling were to collect sample pairs at locations that spanned the expected range of DDT concentration and to take an aliquot for Dexsil Cl⁻ analysis and one for analysis by Method 8080 within a one-foot radius of each other. Figure 4.4 presents the results from these sample pairs.

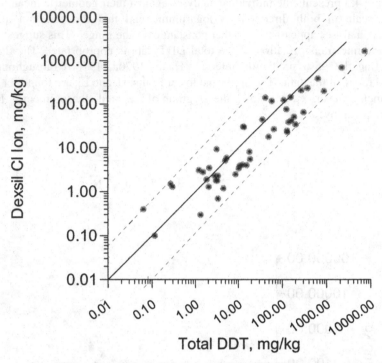

Figure 4.4 Paired Cl Ion versus Total DDT Concentration

Note from this figure that the variation of the data appears to be much the same as that form among replicate Method 8080 analyses. In fact, the dashed lines in Figure 4.4 are exactly the same prediction limits given in Figure 4.3. Therefore, the Dexsil Cl⁻ analysis appears to provide a viable alternative to Method 8080 in measuring the DDT concentration as the paired results from the field sampling experiment appear to be within the measurement precision expected from Method 8080. And, again we use a log-log scale to present the data. This suggests that a log-log model given in Equation [4.22] might be very appropriate for describing the relationship between Dexsil Cl⁻ analysis and the corresponding Method 8080 result for total DDT:

$$\ln(Cl) = \beta_0 + \beta_1 \ln(DDT) \qquad [4.37]$$

Not only does the relationship between the log-transformed Cl⁻ and DDT observations appear to be linear, but the variance of the log-transformed observations appears to be constant over the range of observation. Letting y represent $\ln(Cl^-)$ and x represent $\ln(DDT)$ in Example 4.3 we obtain estimates of β_0 and β_1 via linear least squares.

Fitting the model:

$$y_i = \beta_0 + \beta_1 x_i + \varepsilon_i \qquad [4.38]$$

we obtain estimates of β_0 and β_1 as $\hat{\beta}_0 = 0.190$ and $\hat{\beta}_1 = 0.788$. An important consideration in evaluating the both the statistical and practical significance of these estimates is their correlation. The least squares estimates of the slope and intercept are always correlated unless the mean of the x's is identical to zero. Thus, there is a joint confidence region for the admissible slope-intercept pairs that is elliptical in shape.

Example 4.3 Regression Analysis of Field Calibration Data

Panel 1. The Data

Sample Id.	Cl⁻	y=ln(Cl⁻)	Total DDT	x = ln(DDT)	Sample Id.	Cl⁻	y=ln(Cl⁻)	Total DDT	x = ln(DDT)
SB-001	1.9	0.6419	1.8	0.5988	SB-034	24.4	3.1946	128.6	4.8569
SB-002	2.3	0.8329	3.4	1.2119	SB-034B	43.9	3.7819	35.4	3.5673
SB-005	2.3	0.8329	2.8	1.0296	SB-035	144.2	4.9712	156.2	5.0511
SB-006	22.8	3.1268	130.5	4.8714	SB-036	139.7	4.9395	41.4	3.7233
SB-006	26.5	3.2771	64.9	4.1728	SB-040	30.2	3.4078	12.9	2.5572
SB-007	1653.0	7.4103	7202.0	8.8821	SB-040D	29.7	3.3911	13.5	2.6027
SB-008	34.0	3.5264	201.7	5.3068	SB-046	2.8	1.0296	1.5	0.4114
SB-009	75.6	4.3255	125.0	4.8283	SB-046D	5.1	1.6292	4.9	1.5953
SB-010	686.0	6.5309	2175.0	7.6848	SB-051	0.7	-0.3567	3.4	1.2090
SB-011	232.0	5.4467	370.9	5.9159	SB-054	50.7	3.9259	186.0	5.2257
SB-011D	208.0	5.3375	269.7	5.5973	SB-054D	41.6	3.7281	160.3	5.0770
SB-012	5.5	1.7047	18.6	2.9232	SB-064	0.3	-1.2040	1.3	0.2776
SB-013	38.4	3.6481	140.3	4.9438	SB-066	4.0	1.3863	15.4	2.7344
SB-014	17.8	2.8792	49.0	3.8918	SB-066D	2.5	0.9163	8.6	2.1541
SB-015	1.8	0.5878	3.2	1.1694	SB-069	3.4	1.2238	10.2	2.3224
SB-018	9.3	2.2300	3.1	1.1362	SB-069D	4.1	1.4110	12.4	2.5153
SB-019	64.7	4.1698	303.8	5.7164	SB-084	198.0	5.2883	868.0	6.7662
SS-01	1.8	0.5878	3.0	1.1105	SB-085	3.9	1.3610	10.8	2.3795
SB-014A	384.0	5.9506	635.6	6.4546	SB-088	3.5	1.2528	2.1	0.7467
SB-014AD	123.1	4.8130	109.1	4.6923	SB-090	3.1	1.1314	1.2	0.1906
SB-015A	116.9	4.7613	58.2	4.0639	SB-093	5.9	1.7750	5.3	1.6752
3B-021	0.4	-0.9163	0.1	-2.7646	3B-094	1.3	0.2624	2.0	0.7159
SB-024	0.1	-2.3026	0.1	-2.1628	SB-095	1.5	0.4055	0.3	-1.3209
SB-024D	1.3	0.2624	0.3	-1.2208	SB-096	8.1	2.0919	18.1	2.8943
SB-031B	1.2	0.1823	4.5	1.5019					

Panel 2. The Regression

Linear Regression of ln(Cl⁻) versus ln(DDT)

Predictor Variable	β	Standard Error β	Student's T	P
CONSTANT	0.190	0.184	1.035	0.306
ln(DDT)	0.788	0.048	16.417	<0.0001

R-SQUARED = 0.848

ANOVA Table for Regression

SOURCE	DF	SS	MS	F	P
REGRESSION	1	191.144	181.144	269.525	<0.001

Most elementary statistics text only consider confidence intervals for the slope and intercept separately ignoring their correlation. This leads to the mistaken notion that the joint slope-intercept is rectangular, thus providing a region that is too large. Mandel and Linnig (1957) have provided a procedure for describing the joint elliptical confidence region for the slope and intercept. This yields the region shown in Figure 4.5. The point at the centroid of this 95% confidence ellipse represents the line of best fit.

Note that there is a horizontal and a vertical reference line shown in Figure 4.5. These reference lines represent the intercept and slope expected based upon stoichiometry, assuming all of the Cl⁻ comes from DDT. In other words, if all of the Cl⁻ comes from DDT, then the Cl⁻ concentration would be exactly half the concentration of DDT. Thus in Equation [4.38] the intercept, β_0, would be expected to be $\ln(0.5) = -0.6931$ and the expectation for the slope, β_1, is unity ($\beta_1 = 1$). This conforms to the initial recommendation of the Dexsil distributor, that the concentration of DDT will be twice the Cl⁻ ion concentration.

Note that the elliptical region represents, with 95 percent confidence, the set of all admissible slope-intercept combinations for our calibrating relationship with the least squares estimates as the centroid of the elliptical region. In other words, we are 95 percent confident that the "true" slope-intercept combination is contained within this elliptical region. Because the point of intersection of the reference lines representing the stoichiometric expectation lies outside of the ellipse estimated line is significantly different from the expectation that all of the Cl⁻ comes from DDT.

We must note that the measurements made by the Dexsil® L2000 and the reference Method 8080 are both subject to error. Thus we are in the situation noted earlier in this chapter. The estimate of β_0 and β_1 obtained via linear least squares are biased. We may correct for this in the estimate of β_1 by employing Equation [4.28]. The estimate of σ_u^2 is the S_x^2 from Table 4.3. The estimate of σ_x^2 is obtained from the corrected sum of squares of the log DDT values used in regression. This estimate is

$$\hat{\sigma}_x^2 = \frac{307.876}{49} = 6.2832$$

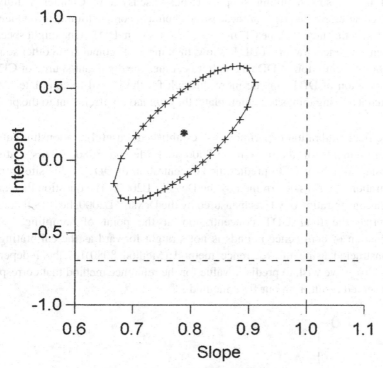

Figure 4.5 **95% Confidence Ellipse,**
Least-Squares Estimates Calibration,
Chloride Ion versus DDT

Applying Equation [4.28] we obtain a maximum likelihood estimate of β_1 as follows:

$$\hat{\beta}_1^* = \hat{\beta}_1 \, \frac{\hat{\sigma}_x^2}{\hat{\sigma}_x^2 - \hat{\sigma}_u^2} = 0.788 \, \frac{6.283}{6.238 - 0.342} = 0.833$$

A maximum likelihood estimate of β_0 can then be obtained by

$$\hat{\beta}_0^* = \bar{y} - \hat{\beta}_1^* \bar{x} = 2.4651 - 0.833 \bullet 2.8874 = 0.059$$

The revised 95% confidence ellipse for the maximum likelihood estimates of the slope and intercept is provided in Figure 4.6.

This confidence ellipse is somewhat closer to the stoichiometric expectation; however, this expectation is still not contained within the elliptical region. If this were a calibration of laboratory instrumentation using certified standards for DDT, we might be concerned over the presence of absolute bias and relative measurement error. However, the relative support of the samples in our field calibration experiment and that employed in the calibration of laboratory instrumentation is entirely different.

All of the soil sampling support issues discussed in Chapter 1 must be considered here. Not the least of these is the chemical composition heterogeneity. It is quite likely that not all of the Cl⁻ in the soil comes from DDT. One might speculate that when the concentration of DDT is low, then most Cl⁻ comes from other sources. When the concentration of DDT is high it becomes the dominant source of Cl⁻, but the dissociation of DDT may be more difficult for the Dexsil to complete. While these are interesting things to contemplate, they are largely irrelevant to the problem at hand.

The field calibration experiment has established a useful relationship between the measurements of the Dexsil® L2000 and Method 8080. This statistical relationship may be used to predict the concentration of DDT at this site from the concentration of Cl⁻ as determined by the Dexsil® L2000. The question then arises, "Given a concentration of Cl⁻ as measured by the Dexsil® L2000, how well does this characterize the total DDT concentration at the point of sampling?" The determination of confidence bounds is not straight forward as the calibrating line was constructed using the reference method (Method 8080) as the independent variable. Now we wish to predict a value for the reference method that corresponds to an observed result from our field method.

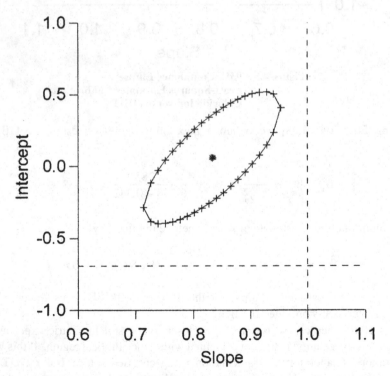

Figure 4.6   95% Confidence Ellipse,
Maximum Likelihood Estimates Calibration,
Chloride Ion versus DDT

The estimated value of the total DDT concentration, x, given an observed concentration of Cl⁻, y_0, is provided by:

$$\hat{x} = \frac{y_0 - \hat{\beta}_0}{\hat{\beta}_1} \qquad [4.39]$$

Mandel and Linnig (1957, 1964) have provided a method for determining reasonable confidence bounds for the predicted value of x in Equation [4.38]. This method starts by considering the following familiar relationship:

$$(y_0 - \hat{\beta}_0 - \hat{\beta}_1 \hat{x})^2 = t^2_{(N-2,\,\alpha/2)} \, S^2 \left[1 + \frac{1}{N} + \frac{(\hat{x} - \bar{x})^2}{\Sigma(x_i - \bar{x})^2} \right] = K^2 \quad [4.40]$$

While this relationship looks intimidating, we only need to calculate K^2 for a given y_0. This gives us the following:

$$\left.\begin{matrix} \hat{x}_U \\[2mm] \hat{x}_L \end{matrix}\right\} = \frac{y_0 - \hat{\beta}_0 \pm K}{\hat{\beta}_1} \qquad [4.41]$$

Repeatedly evaluating [4.40] over a range of y_0's we obtain the confidence limits for an individual total DDT measurement as shown in Figure 4.7.

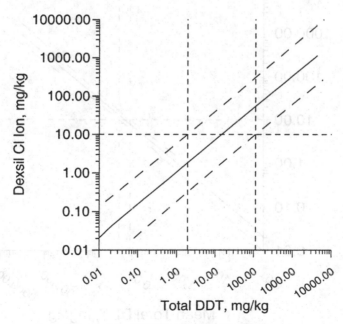

**Figure 4.7   95% Confidence Limits,
Predicted Individual Total DDT**

Note that if we observe a Cl⁻ concentration of 10 mg/kg, then we are 95% confident that if the concentration of total DDT were measured via Method 8080 the result will be between 2 and 112 mg/kg. While this range seems rather large, it may not be unrealistic in light of the sampling variation. Perhaps it is more instructive to consider what the geometric mean Method 8080 concentration might be given an observed a Cl⁻ concentration. Confidence limits for the geometric mean total DDT may be obtained by simply recalculating the value of K as follows:

$$K^2 = t^2_{(N-2,\alpha/2)} S^2 \left[\frac{1}{N} + \frac{(\hat{x} - \bar{x})^2}{\Sigma(x_i - \bar{x})^2} \right]$$

Using this value in Equation [4.40] we obtain Figure 4.8.

Note that given a Cl⁻ concentration of 10 mg/kg we are 95% confident that the mean total DDT concentration for a sampling unit of the size used in the calibration study is between 11 and 20 mg/kg.

Because the Dexsil® L2000 can produce a Cl⁻ analysis in about 15 minutes at a low cost, W. E. Stuck's contractor can cost effectively perform multiple analyses within a proposed exposure/remediation unit. This raises the possibility of near real time guidance of site cleanup efforts.

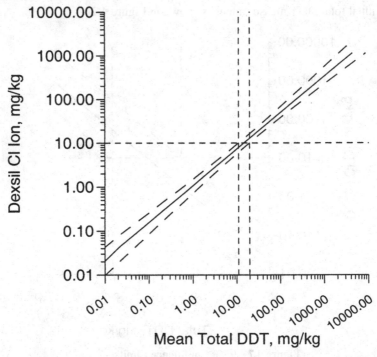

Figure 4.8 95% Confidence Limits for Predicted Main Total DDT

Epilogue

We end this chapter with the wisdom of G. E. P. Box (1966) from his article, "Use and Abuse of Regression" (Box, 1966):

> "In summary the regression analysis of unplanned data is a technique which must be used with great care. However,
>
> (i) It may provide a useful prediction of y in a fixed system being passively observed even when latent variables of some importance exist ....
>
> (ii) It is one of a number of tools sometimes useful in indicating variables which ought to be included in some later planned experiment (in which randomization will, of course, be included as an integral part of the design). It ought never to be used to decide which variable should be excluded from further investigation ....
>
> To find out what happens to a system when you interfere with it you have to interfere with it (not just passively observe it)."[*]

References

Belsley, D. A., Kuh, E., and Welsch., R. E., 1980, *Regression Diagnostics: Identifying Influential Data and Sources of Collinearity*, John Wiley, New York.

Box, G. E. P., 1966, "Use and Abuse of Regression," *Technometrics*, 6(4): 625–629.

Draper, N. R. and Smith, H., 1998, *Applied Regression Analysis*, John Wiley, New York.

Fuller, W. A., 1987, *Measurement Error Models*, New York, Wiley.

Gunst, R. F. and Mason., R. L., 1980, *Regression Analysis and Its Application: A Data-Oriented Approach*, Marcel Dekker, New York.

Gustafson, D. I., and Holden, L. R., 1990, "Nonlinear Pesticide Dissipation in Soil: A New Model Based on Spatial Variability," *Environmental Science and Technology*, 24: 1032–1038.

Hahn, G. J., 1970a, "Statistical Intervals for a Normal Population, Part I, Tables, Examples and Applications," *Journal of Quality Technology*, 2(3): 115–125.

Hahn, G. J., 1970b, "Statistical Intervals for a Normal Population, Part II, Formulas, Assumptions, Some Derivations," *Journal of Quality Technology*, 2(4): 195–206.

LaKind, J. S., Graves, C. G., Ginevan, M. E., Jenkins, R. A., Naiman, D. Q., and Tardiff, R. G., 1999a, "Exposure to Environmental Tobacco Smoke in the Workplace and the Impact of Away-from-Work Exposure," *Risk Analysis*, 19(3): 343–352.

LaKind, J. S., Jenkins, R. A., Naiman, D. Q., Ginevan, M. E., Graves, C. G., and Tardiff, R. G., 1999b, "Use of Environmental Tobacco Smoke Constituents as Markers for Exposure," *Risk Analysis* 19(3): 353–367.

LaKind, J. S., Ginevan, M. E., Naiman, D. Q., James, A. C., Jenkins, R. A., Dourson, M. L., Felter, S. P., Graves, C. G., and Tardiff, R. G., 1999c, "Distribution of Exposure Concentrations and Doses for Constituents of Environmental Tobacco Smoke," *Risk Analysis* 19(3): 369–384.

Lave, L. B. and Seskin, E. P., 1976, "Does Air Pollution Cause Mortality," *Proceedings of the Fourth Symposium on Statistics and the Environment*, American Statistical Association, 1997, pp. 25–35.

Mandel, J. and Linnig, F., J., 1957, "Study of Accuracy in Chemical Analysis Using Linear Calibration Curve," *Analytical Chemistry*, 29: 743–749.

Sokol, R. R. and Rolhf, F. J, 1995, *Biometry*, W. H. Freeman, New York.

Terril, M. E., Ou, K. C., and Splitstone, D. E., 1994, "Case Study: A DDT Field Screening Technique to Guide Soil Remediation," *Proceedings: Ninth Annual Conference on Contaminated Soils*, Amherst Scientific Publishers, Amherst, MA.

Tukey, J. W., 1976, "Discussion of Paper by Lave and Seskin," *Proceedings of the Fourth Symposium on Statistics and the Environment*, American Statistical Association, 1997, pp. 37–41.

USEPA, 1993, *Superfund Innovative Technology Evaluation Program, Technology Profiles Sixth Edition*, EPA/540/R-93/526, pp. 354–355.

Rawlings, J. O. Pantula, S. G., and Dickey, D. A., 1998, *Applied Regression Analysis: A Research Tool*, Springer-Verlag, New York.

CHAPTER 5

Tools for Dealing with Censored Data

"As trace substances are increasingly investigated in soil, air, and water, observations with concentrations below the analytical detection limits are more frequently encountered. 'Less-than' values present a serious interpretation problem for data analysts." (Helsel, 1990a)

Calibration and Analytical Chemistry

All measurement methods (e.g., mass spectrometry) for determining chemical concentrations have statistically defined errors. Typically, these errors are defined as a part of developing the chemical analysis technique for the compound in question, which is termed "calibration" of the method.

In its simplest form calibration consists of mixing a series of solutions that contain the compound of interest in varying concentrations. For example, if we were trying to measure compound A at concentrations of between zero and 50 ppm, we might prepare a solution of A at zero, 1, 10, 20, 40, and 80 ppm, and run these solutions through our analytical technique. Ideally we would run 3 or 4 replicate analyses at each concentration to provide us with a good idea of the precision of our measurements at each concentration. At the end of this exercise we would have a set of N measurements (if we ran 5 concentrations and 3 replicates per concentration, N would equal 15) consisting of a set of k analytic outputs, $A_{i,j}$ for each known concentration, C_i. Figure 5.1 shows a hypothetical set of calibration measurements, with a single A_i for each C_i, along with the regression line that best describes these data.

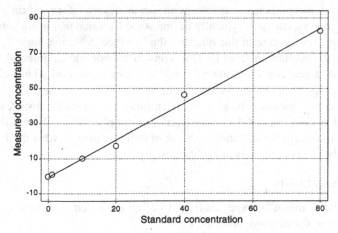

Figure 5.1   A Hypothetical Calibration Curve,
Units are Arbitrary

Regression (see Chapter 4 for a discussion of regression) is the method that is used to predict the estimated measured concentration from the known standard concentration (because the standards were prepared to a known concentration). The result is a prediction equation of the form:

$$M_i = \beta_0 + \beta_1 \cdot C_i + \varepsilon_i \qquad [5.1]$$

Here M_i, is the predicted mean of the measured values (the $A_{i,j}$'s) at known concentration C_i, β_0 the estimated concentration at $C_i = 0$, β_1 is the slope coefficient that predicts M_i from C_i, and ε_i is the error associated with the prediction of M_i.

Unfortunately, Equation [5.1] is not quite what we want for our chemical analysis method because it allows us to predict a measurement from a known standard concentration. When analyses are actually being performed, we wish to use the observed measurement to predict the unknown true concentration. To do this, we must rearrange Equation [5.1] to give:

$$C_i = \frac{M_i - \beta_0}{\beta_1} + \varepsilon'_i \qquad [5.2]$$

In Equation [5.2] β_0 and β_1 are the same as those in [5.1], but C_i is the unknown concentration of the compound of interest, M_i is the measurement from sample i, and ε'_i is the error associated with the "inverse" prediction of C_i from M_i. This procedure is termed inverse prediction because the original regression model was fit to predict M_i from C_i, but then is rearranged to predict C_i from M_i. Note also that the error terms in [5.1] and [5.2] are different because inverse prediction has larger errors than simple prediction of y from x in a regular regression model.

Detection Limits

The point of this discussion is that the reported concentration of any chemical in environmental media is an estimate with some degree of uncertainty. In the calibration process, chemists typically define some C_n value that is not significantly different from zero, and term this quantity the "method detection limit." That is, if we used the ε' distribution from [5.2] to construct a confidence interval for C, C_n would be the largest concentration whose 95% (or other interval width) confidence interval includes zero. Values below the limit of detection are said to be censored because we cannot measure the actual concentration and thus all values less than Cn are reported as "less than LOD," "nondetect," or simply "ND." While this seems a rather simple concept the statistical process of defining exactly what the LOD is for a given analytical procedure is not (Gibbons, 1995).

Quantification Limits

Note that as might be expected from [5.2] all estimated C_i values, $\hat{c}_i$, have an associated error distribution. That is:

$$\hat{c}_i = \kappa_i + \varepsilon_i \qquad [5.3]$$

where κ_i is the true but unknown concentration and ε_i is a random error component. When $\hat{c}_i$ is small, it can have a confidence interval that does not include zero (thus it is not an "ND") but is still quite wide compared to the concentration being reported. For example, one might have a dioxin concentration reported as 500 ppb, but with a 95% confidence interval of 200 to 1,250 ppb. This is quite imprecise and would likely be reported as below the "limit of quantification" or "less than LOQ." However, the fact remains that a value reported as below the limit of quantification still provides evidence that the substance of interest has been identified.

Moreover, if the measured concentrations are unbiased, it is true that the average error is zero. That is:

$$\sum \varepsilon_i = 0 \qquad\qquad [5.4]$$

Thus if we have many values below the LOQ it is true that:

$$\sum \hat{c}_i = \sum \kappa_i + \sum \varepsilon_i \qquad\qquad [5.5]$$

and for large samples,

$$\sum \hat{c}_i = \sum \kappa_i \qquad\qquad [5.6]$$

That is, even if all values are less than LOQ, the sum is still expected to equal the sum of the unknown but true measurements and by extension, the mean of a group of values below the LOQ, but above the DL, would be expected to equal the true sample mean.

It is worthwhile to consider the LOQ in the context of the calibration process. Sometimes an analytic method is calibrated across a rather narrow range of standard concentrations. If one fits a statistical model to such data, the precision of predictions can decline rapidly as one moves away from the range of the data used to fit the model. In this case, one may have artificially high LOQs (and Detection Limit or DLs as well) as a result of the calibration process itself. Moreover, if one moves to concentrations above the range of calibration one can also have unacceptably wide confidence intervals. This leads to the seeming paradox of values that are too large to be acceptably precise. This general problem is an issue of considerable discussion among statisticians engaged in the evaluation of chemical concentration data *(see for example: Gilliom and Helsel, 1986; Helsel and Gilliom, 1986; Helsel, 1990a 1990b).*

The important point to take away from this discussion is that values less than LOQ do contain information and, for most purposes, a good course of action is to simply take the reported values as the actual values (which is our expectation given unbiased measurements). The measurements are not as precise as we would like, but are better than values reported as "<LOQ."

Another point is that sometimes a high LOQ does not reflect any actual limitation of the analytic method and is in fact due to calibration that was performed over a limited range of standard concentrations. In this case it may be possible to improve our understanding of the true precision of the method being used by doing a new calibration study over a wider range of standard concentrations. This will not make our existing <LOQ observations any more precise, but may give us a better idea of how precise such measurements actually are. That is, if we originally had a calibration data set at 200, 400, and 800 ppm and discovered that many field measurements are less than LOQ at 50 ppm, we could ask the analytical chemist to run a new set of calibration standards at say 10, 20, 40, and 80 ppm and see how well the method actually works in the range of concentrations encountered in the environment. If the new calibration exercise suggests that concentrations above 15 ppm are measured with adequate precision and are thus "quantified," we should have greater faith in the precision of our existing less than LOQ observations.

Censored Data

More often, one encounters data in the form of reports where the original raw analytical results are not available and no further laboratory work is possible. Here the data consist of the quantified data that are reported as actual concentrations, the less than LOQ observations that are reported as less than LOQ, together with the concentration defining the LOQ and values below the limit of detection, that are reported as ND, together with concentration defining the limit of detection (LOD). It is also common to have data reported as "not quantified" together with a "quantification limit." Such a limit may reflect the actual LOQ, but may also represent the LOD, or some other cutoff value. In any case the general result is that we have only some of the data quantified, while the rest are defined only by a cutoff value(s). This situation is termed "left censoring" in statistics because observations below the censoring point are on the left side of the distribution.

The first question that arises is: "How do we want to use the censored data set?" If our interest is in estimating the mean and standard deviation of the data, and the number of nonquantified observations (NDs and <LOQs) is low (say 10% of the sample or less), the easiest approach is to simply assume that nondetects are worth 1/2 the detection limit (DL), and that <LOQ values (LVs) are defined as:

$$LV = DL + \tfrac{1}{2}(LOQ - DL) \qquad\qquad [5.7]$$

This convention makes the tacit assumption that the distribution of nondetects is uniformly distributed between the detection limit and zero, and that <LOQ values are uniformly distributed between the DL and the LOQ. After assigning values to all nonquantified observations, we can simply calculate the mean and standard deviation using the usual formulae. This approach is consistent with EPA guidance regarding censored data (e.g., EPA, 1986).

The situation is even easier if we are satisfied with the median and interquartile range as measures of central tendency and dispersion. The median is defined for any data set where more than half of the observations are quantified, while the interquartile range is defined for any data set where at least 75% of the observations are quantified.

Estimating the Mean and Standard Deviation Using Linear Regression

As shown in Chapter 2, observations from a normal distribution tend to fall on a straight line when plotted against their expected normal scores. This is true even if some of the data are below the limit of detection (see Example 5.1). If one calculates a linear regression of the form:

$$C = A + B \cdot Z\text{-Score} \qquad [5.8]$$

where C is the measured concentration, A and B are fitted constants, and Z-Score is the **expected normal score** based on the rank order of the data, A is an estimate of the mean, μ, and B is an estimate of the standard deviation, σ (Gilbert, 1987; Helsel, 1990).

Expected Normal Scores

The first problem in obtaining expected normal scores is to convert the ranks of the data into cumulative percentiles. This is done as follows:

1. The largest value in a sample of N receives rank N, the second largest receives rank $N - 1$, the third largest receives rank $N - 2$ and so on until all measured values have received a rank. In the event that two or more values are tied (in practice this should happen very rarely; if you have many tied values you need to find out why), simply assign one rank K and one rank $K - 1$. For example if the five largest values in a sample are unique, and the next two are tied, assign one rank 6 and one rank 7.

2. Convert each assigned rank, r, to a cumulative percentile, P, using the formula:

$$P = \frac{(r - 3/8)}{(N + 1/4)} \qquad [5.9]$$

We note that other authors (e.g., Gilliom and Helsel, 1986) have used different formulae such as $P = r/(N + 1)$. We have found that using P values calculated using [5.8] provide better approximations to tabled Expected Normal Scores (Rohlf and Sokol, 1969) and thus will yield more accurate regression estimates of μ and σ.

3. Once P values have been calculated for all observations, one can obtain expected normal or Z scores using the relationship:

$$Z(P) = \varphi(P) \qquad [5.10]$$

Here Z(P) is the z-score associated with the cumulative probability P, and ψ is the standard normal inverse cumulative distribution function. This function is shown graphically in Figure 5.2.

4. Once we have obtained Z values for each P, we are ready to perform a regression analysis to obtain estimates of μ and σ.

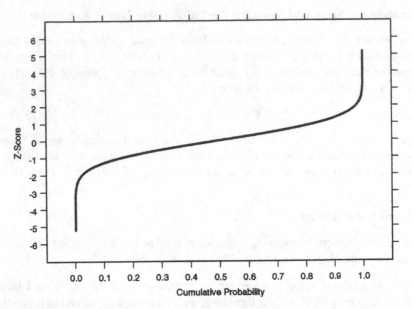

Figure 5.2 The Inverse Normal Cumulative Distribution Function

Example 5.1 contains a sample data set with 20 random numbers, sorted smallest to largest, generated from a standard normal distribution ($\mu = 0$ and $\sigma = 1$), cumulative percentiles calculated from Equation 5.8, and expected normal scores calculated from these P values. When we look at Example 5.1, we see that the estimates for μ and σ look quite close to the usual estimates of μ and σ except for the case where 75% of the data (15 observations) are censored. Note first that even when we have complete data we do not reproduce the parametric values, $\mu = 0$ and $\sigma = 1$. This is because we started with a 20-observation random sample. For the case of 75% censoring the estimated value for μ is quite a bit lower than the sample value of −0.3029 and the estimated value for σ is also a good bit higher than the sample value of 1.0601. However, it is worthwhile to consider that if we did not use the regression method for censored data, we would have to do something else. Let us assume that our detection limit is really 0.32, and assign half of this value, 0.16, to each of the 15 "nondetects" in this example and use the usual formulae to calculate μ and σ. The resulting estimates are $\mu = 0.3692$ and $\sigma = 0.4582$. That is, our estimate for μ is much too large and our estimate for σ is much too small. The moral here is that regression estimates may not do terribly well if a majority of the data is censored, but other methods may do even worse.

The sample regression table in Example 5.1 shows where the **Statistics** presented for the 4 models (20 observations, 15 observations, 10 observations, 5 observations) come from. The CONSTANT term is the intercept for the regression equation and provides our estimate of μ, while the ZSCORE term is the slope of the regression line and provides our estimate of σ. The ANOVA table is included because the regression procedure in many statistical software packages provides this as part of the output. Note that the information required to estimate μ and σ is found in the regression equation itself, not in the ANOVA table. The plot of the data with the regression curve includes both the "detects" and the "nondetects." However, only the former were used to fit the curve. With real data we would have only the detect values, but this plot is meant to show why regression on normal scores works with censored data. That is, if the data are really log-normal, regression on those data points that we can quantify will really describe all of the data. An important point concerning using regression to estimate μ and σ is that all of the tools discussed in our general treatment of regression apply. Thus we can see if factors like influential observations or nonlinearity are affecting our regression model and thus have a better idea of how good our estimates of μ and σ really are.

Maximum Likelihood

There is another way of estimating μ and σ from censored data that also does relatively well when there is considerable left-censoring of the data. This is the method of maximum likelihood. There are some similarities between this method and the regression method just discussed. When using regression we use the ranks of the detected observations to calculate cumulative percentiles and use the standard normal distribution to calculate expected normal scores for the percentiles. We then use the normal scores together with the observed data in a regression model that provides us with estimates of μ and σ. In the maximum likelihood approach we start by assuming a normal distribution for the log-transformed concentration. We then make a guess as to the correct values for μ and σ. Once we have made this guess we can calculate a likelihood for each observed data point, using the guess about μ and σ and the known percentage, ψ, of the data that is censored. We write this result as $L(x_i | \mu, \sigma, \psi)$. Once we have calculated an L for each uncensored observation, we can calculate the overall likelihood of the data, $L(X | \mu, \sigma, \psi)$ as:

$$L(X | \mu, \sigma\psi) = \prod_{i=1}^{N} L(x_i | \mu, \sigma\psi) \qquad [5.11]$$

That is the overall likelihood of the data given μ, σ, and ψ, $L(X | \mu, \sigma, \psi)$, is the product of the likelihoods of the individual data points. Such calculations are usually carried out under logarithmic transformation. Thus most discussions are in terms of log-likelihood, and the overall log-likelihood is the sum of the log-likelihoods of the individual observations. Once $L(X | \mu, \sigma, \psi)$ is calculated there are methods for generating another guess at the values for μ and σ, that yields an even higher log-likelihood. This process continues until we reach values of μ and σ that result in a maximum value for $L(X | \mu, \sigma, \psi)$. Those who want a technical discussion of a representative approach to the likelihood maximization problem in the context of censored data should consult Shumway et al. (1989).

The first point about this procedure is that it is complex compared to the regression method just discussed, and is not easy to implement without special software (e.g., Millard, 1997). The second point is that if there is only one censoring value (e.g., detection limit) maximum likelihood and regression almost always give essentially identical estimates for μ and σ, and when the answers differ somewhat there is no clear basis for preferring one method over the other. Thus for reasons of simplicity we recommend the regression approach.

Multiply Censored Data

There is one situation where maximum likelihood methods offer a distinct advantage over regression. In some situations we may have multiple "batches" of data that all have values at which the data is censored. For example, we might have a very large environmental survey where the samples were split among several labs that had somewhat different instrumentation and thus different detection and quantification limits. Alternatively, we might have samples with differing levels of "interference" for the compound of interest by other compounds and thus differing limits for detection and quantification. We might even have replicate analyses over time with declining limits of detection caused by improved analytic techniques. The cause does not really matter, but the result is always a set of measurements consisting of several groups, each of which has its own censoring level.

One simple approach to this problem is to declare all values below the highest censoring point (the largest value reported as not quantified across all groups) as censored and then apply the regression methods discussed earlier. If this results in minimal data loss (say, 5% to 10% of quantified observations), it is arguably the correct course. However, in some cases, especially if one group has a high censoring level, the loss of quantified data points may be much higher (we have seen situations where this can exceed 50%). In such a case, one can use maximum likelihood methods for multiply censored data such as those contained in Millard (1997) to obtain estimates for μ and σ that utilize all of the available data. However, we caution that estimation in the case of multiple censoring is a complex issue. For example, the pattern of censoring can affect how one decides to deal with the data. When dealing with such complex issues, we strongly recommend that a professional statistician, one who is familiar with this problem area, be consulted.

Example 5.1

The Data for Regression

Y Data (Random Normal) Sorted Smallest to Largest	Cumulative Proportion from Equation 5.8	Z-Scores from Cumulative Proportions
−2.012903	0.030864	−1.868241
−1.920049	0.080247	−1.403411
−1.878268	0.129630	−1.128143
−1.355415	0.179012	−0.919135
−0.986497	0.228395	−0.744142
−0.955287	0.277778	−0.589455
−0.854412	0.327161	−0.447767
−0.728491	0.376543	−0.314572
−0.508235	0.425926	−0.186756
−0.388784	0.475307	−0.061931

The Data for Regression (Cont'd)

Y Data (Random Normal) Sorted Smallest to Largest	Cumulative Proportion from Equation 5.8	Z-Scores from Cumulative Proportions
−0.168521	0.524691	0.061932
0.071745	0.574074	0.186756
0.084101	0.623457	0.314572
0.256237	0.672840	0.447768
0.301572	0.722222	0.589456
0.440684	0.771605	0.744143
0.652699	0.820988	0.919135
0.694994	0.870370	1.128143
1.352276	0.919753	1.403412
1.843618	0.969136	1.868242

Statistics

- Summary Statistics for the Complete y Data, using the usual estimators:

  Mean = −0.3029   SD = 1.0601

- Summary Statistics for the Complete Data, using regression of the complete data on Z- Scores:

  Mean = −0.3030   SD = 1.0902 $R^2 = 0.982$

- Summary Statistics for the 15 largest y observations (y = -0.955287 and larger), using regression of the data on Z- Scores:

  Mean = −0.3088   SD = 1.1094 $R^2 = 0.984$

- Summary Statistics for the **10 largest y observations** (y = −0.168521 and larger), using regression of the data on Z- Scores:

  Mean = −0.2641   SD = 1.0661 $R^2 = 0.964$

- Summary Statistics for the 5 largest y observations (y = 0.440684 and larger), using regression of the data on Z- Scores:

  Mean = −0.5754.   SD = 1.2966 $R^2 = 0.961$

The Regression Table and Plot for the 10 Largest Observations

Unweighted Least-Squares Linear Regression of Y

Predictor Variables	Coefficient	Std Error	Student's t	P
Constant	−0.264	0.068	−3.87	0.0048
Z-score	1.066	0.074	14.66	0.0000

ANOVA Table

Source	DF	SS	MS	F	P
Regression	1	3.34601	3.34601	214.85	0.0000
Residual	8	0.12459	0.01557		
Total	9	3.47060			

R-SQUARED   0.9641

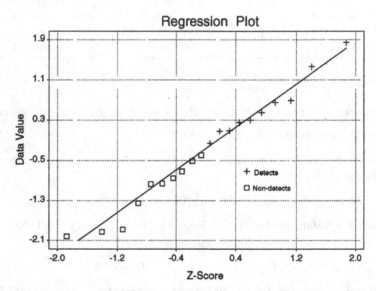

Figure 5.3 A Regression Plot of the Data Used in Example 5.1

Estimating the Arithmetic Mean and Upper Bounds on the Arithmetic Mean

In Chapter 2, we discussed how one can estimate the arithmetic mean concentration of a compound in environmental media, and how one might calculate an upper bound on this arithmetic mean. Our general recommendation was to use the usual statistical estimator for the arithmetic mean and to use bootstrap methodology (Chapter 6) to calculate an upper bound on this mean. The question at hand is how do we develop estimates for the arithmetic mean, and upper bounds for this mean, when the data are censored?

One approach that is appealing in its simplicity is to use the values of μ and σ, estimated by regression on expected normal scores, to assign values to the censored observations. That is, if we have N observations, k of which are censored, we can assume that there are no tied values and that the ranks of the censored observations are 1 through k. We can then use these ranks to calculate P values using Equation [5.9], and use the estimates P values to calculate expected normal scores

(Equation [5.10]). We then use the regression estimates of μ and σ to calculate "values" for the censored observations and use an exponential transformation to calculate observations in original units (usually ppm or ppb). Finally, we use the "complete" data, which consists of estimated values for the censored observations and observed values for the uncensored observations, together with the usual formulae to calculate $\bar{x}$ and s.

Consider Example 5.2. The estimates of μ and σ are essentially identical. What is perhaps more surprising is the fact that the upper percentiles of the bootstrap distribution shown Example 5.2 are also virtually identical for the complete and partially estimated exponentially transformed data. Replacing the censored data with their exponentially transformed expectations from the regression model and then calculating and s using the resulting pseudo-complete data is a strategy that has been recommended by other authors (Helsel, 1990b; Gilliom and Helsel, 1986; Helsel and Gilliom, 1986). The use of the same data to estimate an upper bound for $\bar{x}$ is a relatively new idea, but one that flows logically from previous work. That is, the use of the bootstrap technique to estimate an upper bound on $\bar{x}$ is well established for the case of uncensored data. As noted earlier (Chapter 2), environmental data is almost always skewed to the right. That is, the distribution has a long "tail" that points to the right. Except for cases of extreme censoring, this long tail always consists of actual observations, and it is this long tail that plays the major role in determining the bootstrap upper bound on $\bar{x}$. Our work suggests that the bootstrap is a useful tool for determining an upper bound on $\bar{x}$ whenever at least 50% of the data are uncensored (Ginevan and Splitstone, 2002).

Example 5.2

Calculating the Arithmetic Mean and its Bootstrap Upper Bound

Y Data (Random Normal) Sorted Smallest to Largest	Z-Scores from Cumulative Proportions	Data Calculated from Estimates of μ and σ	Exponential Transform of Calculated for Censored and Observed for Uncensored
Censored	−1.868240	−2.255831	0.1047864
	−1.403411	−1.760276	0.1719973
	−1.128143	−1.466813	0.2306594
	−0.919135	−1.243989	0.2882319
	−0.744142	−1.057429	0.3473474
	−0.589455	−0.892518	0.4096230
	−0.447767	−0.741464	0.4764157
	−0.314572	−0.599465	0.5491052
	−0.186756	−0.463200	0.6292664
	−0.061931	−0.330124	0.7188341

Calculating the Arithmetic Mean and its Bootstrap Upper Bound (Cont'd)

Y Data (Random Normal) Sorted Smallest to Largest	Z-Scores from Cumulative Proportions	Data Calculated from Estimates of μ and σ	Exponential Transform of Calculated for Censored and Observed for Uncensored
−0.168521	0.061932		0.8449135
0.071745	0.186756		1.0743813
0.084101	0.314572		1.0877387
0.256237	0.447768		1.2920589
0.301572	0.589456	Observed	1.3519825
0.440684	0.744143		1.5537696
0.652699	0.919135		1.9207179
0.694994	1.128143		2.0036971
1.352276	1.403412		3.8662150
1.843618	1.868242		6.3193604

Statistics

- Summary statistics for the complete exponentially transformed Y data from Example 5.1 (column 1), using the usual estimators:

    Mean = 1.2475 SD = 1.4881

- Summary statistics for the exponentially transformed Y data from column 4 above:

    Mean = 1.2621 SD = 1.4797

- Bootstrap percentiles (2,000 replications) for the exponentially transformed complete data from Example 5.1 and from column 4 of Example 5.2.

	50%	75%	90%	95%
Example 5.1	1.2283	1.4501	1.6673	1.8217
Example 5.2	1.2446	1.4757	1.7019	1.8399

Zero Modified Data

The next topic we consider in our discussion of censored data is the case referred to as zero modified data. In this case a certain percentage, Z%, of the data are true zeros. That is, if we are interested in pesticide residues on raw agricultural commodities, it may be that Z% of the crop was not treated with pesticide at all and thus has zero residues. Similarly, if we are sampling groundwater for contamination, it may be that Z% of the samples represent uncontaminated wells and are thus true zeros. In many cases, we have information on what Z% might be. That is, we might know that approximately 40% of the crop was untreated or that 30% of the wells are uncontaminated.

In such a case the expected proportion of samples with any residues θ (both above and below the censoring limit(s)) is:

$$\theta = 1 - (Z\%/100) \qquad [5.12]$$

That is, if we have N samples, we would expect about $L = N \bullet \theta$ samples (L is rounded to the nearest whole number) to have residues.

One simple, and reasonable, way to deal with true zeros is to assume a value for Z%, calculate the number, L, of observations that we expect to have residues, and then use L and O, the number of observations that have observed residues to calculate regression estimates for μ and σ. That is, we assume that we have a sample of size L, with O samples with observed residues. We then calculate percentiles and expected normal scores assuming a sample of size L and proceed as in Example 5.1. In this simple paradigm we could also estimate the L-zero values with undetected residues using the approach shown in Example 5.2 by assigning regression estimated values to these observations. We could then exponentially transform the values for "contaminated" samples to get concentrations in original units, assign the value zero to the N-L uncontaminated samples and use the usual estimator to calculate mean contamination and the bootstrap to calculate an upper bound on the mean.

If we have a quite good idea of Z% and a fairly large sample (say, an L value of 30 or more with at least 15 samples with measured residues), this simple approach is probably all we need, but in some cases we have an idea that Z% is not zero, but are not really sure how large it is. Here one possibility is to use maximum likelihood methods to estimate μ, σ and Z%. Likewise we could also assume a distribution reflecting our uncertainty about Z% (e.g., say we assume that Z% is uniformly distributed between 10 and 50) and use Monte Carlo simulation methods to calculate an uncertainty distribution for the mean. In practice, such approaches may be useful, but both are beyond the scope of this discussion. We have again reached the point at which a subject matter expert should be consulted.

Completely Censored Data

Sometimes we have a large number of observations with no detected concentrations. Here it is common to assign a value of 1/2 the LOD to all observations. This can cause problems because the purpose of risk assessment one often calculates a hazard index (HI). The Hazard Index (HI) for N chemicals is calculated as (EPA, 1986):

$$HI = \left[\sum_{i=1}^{N} \frac{RfD_i}{E_i} \right]^{-1} \qquad [5.13]$$

where the RfD_i is the reference dose, or level below which no adverse effect is expected for the ith chemical compound, and E_i is the exposure expected from that chemical. A site with an HI of greater than one is assumed to present undue risks to human health. If the number of chemicals is large and/or the LODs for the chemicals are high, one can have a situation where the HI is above 1 for a site where no hazardous chemicals have been detected!

The solution to this dilemma is to remember (Chapter 2) that if we have N observations, we can calculate the median cumulative probability for the largest sample observation, P(max), as:

$$P(max) = (0.5)^{1/N} \qquad [5.14]$$

For the specific case of a log-normal distribution, S_P, the number of logarithmic standard deviation error (σ) units that are between the cumulative probability, P(max) of the distribution, and the mean of the parent distribution, is found as the Normal Inverse, Z_I, of P(max), that is:

$$S_P - Z_I[P(max)] = \begin{pmatrix} \text{that standard normal deviate} \\ \text{corresponding to the cumulative} \\ \text{probability, P(max)} \end{pmatrix} \qquad [5.15]$$

To get an estimate of the logarithmic mean, μ, of the log-normal distribution the X_P value, together with the LSE estimate and the natural logarithm of the LOD, LN(LOD), are used:

$$\mu = Ln(LOD) - S_P\sigma \qquad [5.16]$$

The geometric mean, GM, is given by:

$$GM = e^\mu \qquad [5.17]$$

Note that quantity of interest for health risk calculations is often the arithmetic mean, M, which can be calculated as:

$$M = e^{\left(\mu + \frac{\sigma^2}{2}\right)} \qquad [5.18]$$

(see Gilbert, 1987).

We can easily obtain P(max), but how can we estimate σ? In general, environmental contaminants are chemicals dissolved in a matrix (water, soil, peanut butter). To the extent that the same forces operate to vary concentrations, the variations tend to be multiplicative (e.g., if the volume of solvent doubles, the concentrations of all solutes are halved). On a log scale this means that, in the same matrix, high-concentration compounds should have an LSE that is similar to the LSE of low-concentration compounds, because both have been subjected to a similar series of multiplicative concentration changes. Thus we can estimate σ by assuming it is similar to the observed σ values of other compounds with large numbers of detected values. Of course, when deriving an LSE in this manner, one should restrict consideration to chemically similar pairs of compounds (e.g., metal oxides; polycyclic aromatic hydrocarbons). Nonetheless, the σ value of calcium in groundwater might be a useful approximation for the σ of cadmium in groundwater. This approach, together with defensibly conservative assumptions, could be used to estimate a σ for almost any pollutant or food contaminant. Moreover, we need not restrict ourselves to a single σ estimate; we could try a range of values to evaluate the sensitivity of our estimate for M. The procedure discussed here is presented in more detail in Ginevan (1993). An example calculation is shown in Example 5.3.

Note also that if one can calculate a lower bound for P(max). That is, if one wants a 90% lower bound for P(max) one uses 0.10 instead of 0.50 in Equation [5.14]; similarly, if one wants a 95% lower bound one uses 0.05. More generally, if one wants a $1 - \alpha$ lower bound on P(max), one uses α instead of 0.5. This approach may be useful because using a lower bound on P(max) will give an upper bound on μ, which may be used to ensure a "conservative" (higher) estimate for the GM.

Example 5.3

Bounding the mean when all observations are below the LOD:

1. Assume we have a σ value of 1 (experience suggests that many environmental contaminants have σ between 0.7 and 1.7) and a sample size, N, of 200. Also assume that the LOD is 1 part per billion (1 ppb).

2. To estimate a median value for the geometric mean we use the relationship:

$$P(max) = 0.5^{1/200}$$

Thus, P(max) = 0.99654.

3. We now determine from [5.15] that $S_P = Z_I (0.99654) = 2.7007$.

4. The estimate for the logarithmic mean, μ, is given by [5.16] and is:

$$\mu = Ln(LOD) - S_P \bullet \sigma$$

$$\mu = Ln(1) - (2.7007 \bullet 1)$$

$$\mu = -2.7007$$

5. Using [5.17] the estimate for the geometric mean, GM, is:

$$GM = e^\mu$$

$$GM = e^{-2.7007}$$

$$GM = 0.06716$$

6. We can also get an estimate for the arithmetic mean from [5.18] as:

$$M = e^{\left(\mu + \frac{\sigma^2}{2}\right)}$$

$$M = e^{-2.7002 + 1/2}$$

$$M = 0.1107$$

Note that even the estimated arithmetic mean is almost 5-fold less than the default estimate of 1/2 the LOD or 0.5.

When All Else Fails

Compliance testing presents yet another set of problems in dealing with censored data. In many respects this is a simpler problem in that a numerical estimate of average concentration is not necessarily required. However, this problem is perhaps a much more common dilemma than the assessment of exposure risk. Ultimately all one needs to do is demonstrate compliance with some standard of performance within some statistical certainty.

The New Process Refining Company has just updated its facility in Gosh Knows Where, Ohio. As a part of this facility upgrade, New Process has installed a new Solid Waste Management Unit (SWMU), which will receive some still bottom sludge. The monitoring of quality of groundwater around this unit is required under their permit to operate.

Seven monitoring wells have been appropriately installed in the area of the SWMU. Two of these wells are thought to be up gradient and the remaining five down gradient of the SWMU. These wells have been sampled quarterly for the first year after installation to establish site-specific "background" groundwater quality.

Among the principal analytes for which monitoring is required is Xylene. The contract-specified MDL for the analyte is 10 microgram per liter (μg/L). All of the analytical results for Xylene are reported as below the MDL. Thus, one is faced with characterizing the background concentrations of Xylene in a statistically meaningful way with all 28 observations reported as <10 μg/L.

One possibility is to estimate the true proportion of "background" Xylene measurements that can be expected to be above the MDL. This proportion must be somewhere within interval 0.0, and 1.0. Here 0.0 indicates that Xylene will NEVER be observed above the MDL and 1.0 indicates that Xylene will ALWAYS be observed above the MDL. The latter is obviously not correct based upon the existing evidence. While the former is a possibility, it is unlikely that a Xylene concentration will never be reported above the MDL with continued monitoring of background water quality.

There are several reasons why we should consider the likelihood of a future background groundwater Xylene concentration reported above the MDL. Some of these are related to random fluctuations in the analytical and sampling techniques employed. A major reason for expecting a future detected Xylene concentration is that the New Process Refining Company facility lies on top of a known petroleum-bearing formation. Xylenes occur naturally in such formations (Waples, 1985).

Fiducial Limits

Thus, the true proportion of Xylene observations possibly above the MDL is not well characterized by the point estimate, 0.0, derived from the available evidence. This proportion is more logically something greater than 0.0, but certainly not 1.0. We may bound this true proportion by answering the question: "What are possible values of the true proportion, p, of Xylene observations greater than the MDL which would likely have generated the available evidence?"

First, we need to define "likely" and then find a relationship between this definition and p. We can define a "likely" interval for p as those values of p that could have generated the current evidence with 95 percent confidence (i.e., a probability of 0.95). Since there are only two alternatives, either a concentration value is above, or it is below, the MDL, the binomial density model introduced in Equation [2.23] provides a useful link between p and the degree of confidence.

The lowest possible value of p is 0.0. As discussed in the preceding paragraphs, if the probability of observing a value greater than the MDL is 0.0, then the sample results would occur with certainty. The upper bound, p_u, of our set of possible values for p will be the value that will produce the evidence with a probability of 0.05. In other words we are 95 percent confident that p is less than this value. Using Equation [2.23], this is formalized as follows:

$$f(x=0) = \binom{28}{0} P_u^0 (1 - p_u)^{28} \geq 0.05$$

Solving for p_u, [5.13]

$$p_u = 1.0 - (0.05)^{1/28} = 0.10$$

The interval $0.0 \leq p \leq 0.10$ not only contains the "true" value of p with 95 percent confidence, it is also a "fiducial interval." Wang (2000) provides a nice discussion of fiducial intervals including something of their history. Fiducial intervals for the binomial parameter p were proposed by Clopper and Pearson in 1934.

The construction of a fiducial interval for the probability of getting an observed concentration greater than the MDL is rather easy when all of the available observations are below the MDL. However, suppose one of our 28 "background" groundwater quality observations is above the MDL. Obviously, this eliminates 0.0 as a possible value for the lower bound.

We may still find a fiducial interval, $p_L \leq p \leq p_U$, by finding the bounding values that satisfy the following relations:

$$\text{Prob}(x < X | p_L) = \alpha / 2$$

$$[5.14]$$

$$\text{Prob}(x > X | p_U) = \alpha / 2$$

Here $(1 - \alpha)$ designates the desired degree of confidence, X represents the observed number of values exceeding the MDL. Slightly rewriting [5.14] as follows, we may use the identity connecting the beta distribution and the binomial distribution to obtain values for p_L and p_U (see Guttman, 1970):

$$\text{Prob}(x < X | p_L) = \alpha / 2$$

$$[5.15]$$

$$\text{Prob}(x \leq X | p_U) = \alpha / 2$$

In the hypothetical case of one out of 28 observations reported as above the MDL, $p_L = 0.0087$ and $p_U = 0.1835$. Therefore the 95 percent fiducial, or confidence, interval (0.0087, 0.1835) for p.

The Next Monitoring Event

Returning to the example provided by New Process Refining Company, we have now bounded the probability that a Xylene concentration above the MDL will be observed. The fiducial interval for this probability based upon the background monitoring event is (0.0, 0.10). One now needs to address the question of when there should be concern that groundwater quality has drifted from background. If on the next monitoring event composed of a single sample from each of the seven wells, one Xylene concentration was reported as above the MDL would that be cause for concern? What about two above the MDL? Or perhaps three?

Christman (1991) presents a simple statistical test procedure to determine whether or not one needs to be concerned about observations greater than the MDL. This procedure determines the minimum number of currently observed monitoring results reported as above the MDL that will result in concluding there is a potential problem, while controlling the magnitude of the Type I and Type II decision errors.

The minimum number of currently observed monitoring results reported as above the MDL will be referred to as the "critical count" for brevity. We will represent the critical count by "K." The Type I error is simply the probability of observing K or more monitoring results above the MDL on the next round of groundwater monitoring given the true value of p is within the fiducial interval:

$$\text{Prob}(\text{Type I Error} | K) = 1.0 - \sum_{k=0}^{K-1} \binom{7}{k} p^k (1.0 - p)^{7-k} \quad [5.16]$$

where

$$(0.0 \leq p \leq 0.1)$$

This relationship is illustrated in Figure 5.4.

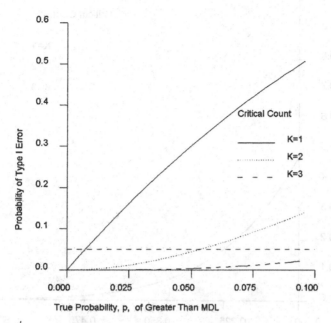

**Figure 5.4   Probability of Type I Error
for Various Critical Counts**

If one were to decide that a possible groundwater problem exists based upon one exceedance of the MDL in the next monitoring event, i.e., a critical count of 1, the risk of falsely reaching such a conclusion dramatically increases to nearly 50 percent as the true value of p approaches 0.10, the upper limit of the fiducial interval. If we choose a critical count of 3, the risk of falsely concluding a problem exists remains at less than 0.05 (5 percent).

Fixing the Type I error is only part of the equation in choosing an appropriate critical count. Consistent with Steps 5, 6, and 7 of the Data Quality Objects Process (USEPA, 1994), one needs to consider the risk of falsely concluding that no groundwater problem exists when in fact p has exceeded the upper fiducial limit. This is the risk of making a decision error of Type II. The probability of a Type II error is easily determined via Equation [5.17]:

$$\text{Prob}(\text{Type II Error}|K) = 1.0 - \sum_{k=0}^{K-1} \binom{7}{k} p^k (1.0 - p)^{7-k} \quad [5.17]$$

where

$$(0.1 < p)$$

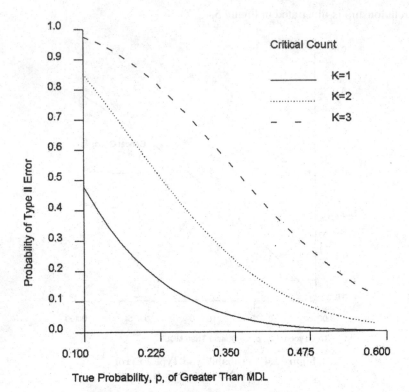

**Figure 5.5   Probability of Type II Errors
for Various Critical Counts**

Note that the risk of making a Type II error is near 90 percent for a critical count of 3, Prob(Type II Error|K = 3) ≥ 0.90, when p is near 0.10 and remains greater than 20 percent for values of p near 0.5. Therefore, while a critical count of 3 minimizes the operator's risk of a falsely concluding a problem may exist (Type I error) the risk of falsely concluding no problem exists (Type II error) remains quite large.

Suppose that a critical count of two seems reasonable, what are the implications for the groundwater quality decision making? New Process Refining Company must be willing to run a greater than 5 percent chance of a false allegation of groundwater quality degradation if the true p is between 0.05 and 0.10. Conversely, the other stakeholders must take a greater than 20 percent chance that no degradation of quality will be found when the true p is between 0.10 and 0.37. This interval of 0.05 ≤ p ≤ 0.37 is often referred to as the "gray of region" (USEPA, 1994, pp. 34–36). This is a time for compromise and negotiation.

Epilogue

There is no universal tool to use in dealing with censored data. The tool one chooses to use depends upon the decision one is attempting to make and the consequences associated with making an incorrect decision. Even then there may be several tools that can accomplish the same task. The choice among them depends largely on the assumptions one is willing to make. As with all statistical tools, the choice of the best tool for the job depends upon the appropriateness of the underlying assumptions and the recognition and balancing of the risks of making an incorrect decision.

References

Christman, J. D., 1991, "Monitoring Groundwater Below Limits of Detection," *Pollution Engineering*, January.

Clopper, C. J. and Pearson, E. S., 1934, "The Use of Confidence or Fiducial Limits Illustrated in the Case of the Binomial," *Biometrika*, 26, 404–413.

Environmental Protection Agency (EPA), 1986, *Guidelines for the Health Risk Assessment of Chemical Mixtures*, 51 FR 34014-34025.

Gibbons, R. D., 1995, "Some Statistical and Conceptual Issues in the Detection of Low Level Environmental Pollutants," *Environmental and Ecological Statistics* 2: 125–144.

Gilbert, R. O., 1987, *Statistical Methods for Environmental Pollution Monitoring*, Van Nostrand Reinhold, New York.

Gilliom, R. J. and Helsel, D. R., 1986, "Estimation of Distributional Parameters for Censored Trace Level Water Quality Data 1: Estimation Techniques," Water Resources Research, 22: 135–146.

Ginevan, M. E., 1993, "Bounding the Mean Concentration for Environmental Contaminants When all Observations are below the Limit of Detection," American Statistical Association, 1993, *Proceedings of the Section on Statistics and the Environment*, pp. 123–128.

Ginevan, M. E. and Splitstone, D. E., 2001, "Bootstrap Upper Bounds for the Arithmetic Mean of Right-Skewed Data, and the Use of Censored Data," *Environmetrics* (in press).

Guttman, I., 1970, *Statistical Tolerance Regions: Classical and Bayesian*, Hafner Publishing Co., Darien, CT.

Helsel, D. R. and Gilliom, R. J., 1986, "Estimation of Distributional Parameters for Censored Trace Level Water Quality Data 2: Verification and Applications," *Water Resources Research*, 22: 147–155.

Helsel, D. R., 1990a, "Statistical Analysis of Data Below the Detection Limit: What Have We Learned?, Environmental Monitoring, Restoration, and Assessment: What Have We Learned?", *Twenty-Eighth Hanford Symposium on Health and the Environment*, October 16–19, 1989, ed. R.H. Gray, Battelle Press, Columbus, OH.

Helsel, D. R., 1990b, "Less Than Obvious: Statistical Treatment of Data below the Detection Limit," *Environmental Science and Technology*, 24: 1766–1774.

Millard, S. P., 1997, *Environmental Stats for S-Plus*. Probability, Statistics and Information, Seattle, WA.

Rohlf, F. J. and Sokol, R. R., 1969, *Statistical Tables*, Table AA, W. H. Freeman, San Francisco.

Shumway, R. H., Azari, A. S., and Johnson, P., 1989, "Estimating Mean Concentrations Under Transformation for Environmental Data with Detection Limits," *Technometrics*, 31: 347–356.

USEPA, 1994, *Guidance for the Data Quality Objectives Process*, EPA QA/G-4.

Wang, Y. H., 2000, "Fiducial Intervals: What Are They?," *The American Statistician*, 52(2): 105–111.

Waples, D. W., 1985, *Geochemistry in Petroleum Exploration*, Reidl Publishing, Holland.

CHAPTER 6

The Promise of the Bootstrap

"A much more serious fallacy appears to be involved in Galton's assumption that the value of the data, for the purpose for which they were intended, could be increased by rearranging the comparisons. Modern statisticians are familiar with the notions that any finite body of data contains only a limited amount of information, on any point under examination; that this limit is set be the nature of the data themselves, and cannot be increased by any amount of ingenuity expended in their statistical examination: that the statistician's task, in fact, is limited to the extraction of the whole of the available information on any particular issue. If the results of an experiment, as obtained, are in fact irregular, this evidently detracts from their value; and the statistician is not elucidating but falsifying the facts, who rearranges them so as to give an artificial appearance of regularity." (Fisher, 1966)

Introductory Remarks

The wisdom of Fisher's critique of Francis Galton's analysis of data from Charles Darwin's experiment on plant growth holds true today for the analysis of environmental data as it was when penned in 1935 in regard to experimental design in the biological sciences. The point is that a given set of data collected for a specific purpose contains only a limited amount of information regarding the population from which they were obtained. This limited amount of information is set by the data themselves and the manner in which they were obtained. No amount of ingenuity on the part of the data analyst can increase that amount of information.

In order for the information contained in any set of data to be useful, one must assume that the data at hand are *representative* of the entity about which we desire information. If we desire to assess the risk of an individual moving around a residential lot, then we must assume that the soil samples used to assess the analyte concentration on that lot truly *represent* the concentrations to which an individual might possibly, and reasonably, be exposed. This assumption is basic to making any inference regarding the analyte concentration for the residential lot.

Fisher's comments must also be interpreted in their historical context. The Gaussian "theory of errors" and the work of "Student" published in 1908 were relatively new ideas in 1935. These ideas provide convenient and efficient means for extracting information and "making sense" out of the data and are widely taught in basic and advanced courses on statistics. However, the Gaussian model may not always be useful in extracting the data's information. This is particularly true for environmental studies where the data distributions may be highly skewed.

Here we disagree slightly with Fisher. The *irregularity* of environmental data does not detract from its utility but often provides the only really useful information.

Those who attempt to squeeze these data into some convenient assumed model are indeed often "falsifying the facts" to meet the demands of an assumed model that may not be appropriate. The recent advent of powerful and convenient computing equipment limits the requirement of using well-studied and convenient models for information extraction. The nonparametric technique known as the "bootstrap" (Efron and Tibshirani, 1993) provides great promise for estimating parameters and confidence bounds of interest to environmental scientists and risk assessors, and even in testing statistical hypotheses.

We will use four examples to illustrate the efficacy of bootstrap resampling. The first will consider estimation of the 95% upper confidence limit (UCL) on the average exposure to arsenic of a person randomly moving around a residential lot. The second example will take up the problems of estimating a daily effluent discharge limit appropriate for obtaining a waste water discharge permit under the National Pollution Discharge Elimination System (NPDES) as required by the Clean Water Act. Third, we will consider the problem of estimation of the ratio of uranium 238 (U_{238}) to radium 226 (Ra_{226}) for use in the determination of the concentration U_{238} in site soil using sodium iodide gamma spectroscopy, which is not capable of measuring U_{238} directly. Lastly, we propose a bootstrap alternative to the two sample t-test.

The assumptions underlying the bootstrap will be given particular attention in our discussion. The reader will note that the required assumptions are a subset of those underlying the application of most parametric statistical techniques.

The Empirical Cumulative Distribution

Located near the Denver Stapleton Airport is a Superfund site known as the "Vasquez Boulevard and I-70 Site." One of the reasons this site is interesting is that Region 8 of the USEPA (ISSI, 1999) in cooperation with the City and County of Denver and the Colorado Department of Public Health and Environment have conducted extensive sampling of surface soil (0–2 inches) on a number of residential properties. Figure 6.1 presents the schematic map of sampling locations at Site #3 in this study. Two hundred and twenty-four (224) samples were collected on a grid with nominal 5-foot spacing.

Figure 6.2 presents the histogram of arsenic concentration in the surface soil of this residential site as represented by the collected samples. Note that the concentration scale is logarithmic. The astute reader will also note the suggestion that the data do not appear to conform to any "classical" statistical model.

ASSUMING that these data are representative of the arsenic concentration on the lot, we might be interested in the proportion of the soil concentrations that are below some fixed level. To that end we may construct a cumulative histogram as illustrated in Figure 6.3 simply by stacking the histogram bars on top of one another from left to right. Connecting the upper right-hand corner of each stacked bar, we have a representation of the empirical cumulative distribution function (ecdf). A smoother and more formal rendition of the ecdf can be obtained by arranging the data in order by increasing concentration and plotting the data values versus their relative ranks.

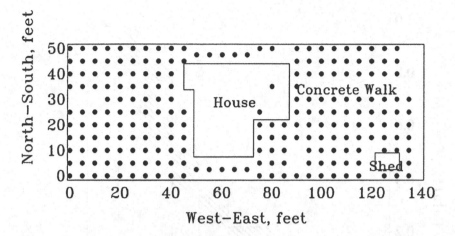

Figure 6.1 Sampling Locations,
Residential Risk-Based Sampling Site #3 Schematic,
Vasquez Boulevard and I-70 Site, Denver, CO

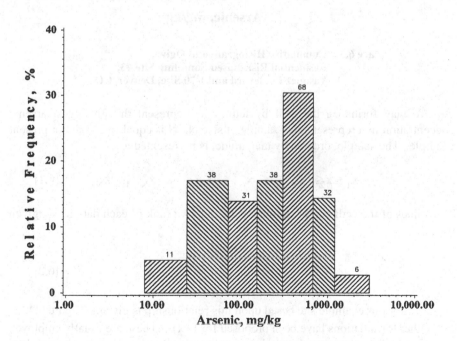

Figure 6.2 Histogram of Arsenic Concentrations,
Residential Risk-Based Sampling Site #3,
Vasquez Boulevard and I-70 Site, Denver, CO

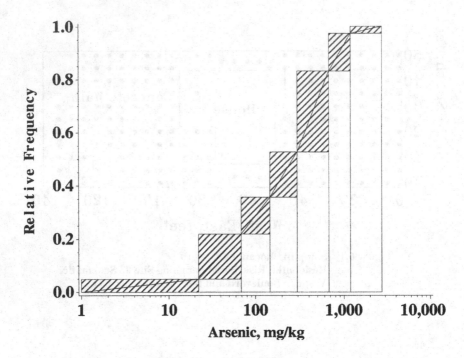

Figure 6.3   **Cumulative Histogram and Ogive,**
                Residential Risk-Based Sampling Site #3,
                Vasquez Boulevard and I-70 Site, Denver, CO

We may formalize the ecdf by letting x_i represent the ith largest arsenic concentration in a representative sample of size N. N is equal to 224 in our present example. The sample, ordered by magnitude, is represented as:

$$x_1 \leq x_2 \leq x_3 \leq . \leq x_i \ ... \leq x_{N-2} \leq x_{N-1} \ x_N \qquad [6.1]$$

The values of the ecdf can be defined as the relative rank of each datum, x_i, given by:

$$F(x_i) = \frac{i}{N} \qquad [6.2]$$

The ecdf for our example site based upon this relationship is given in Figure 6.4.

Other formulations have been proposed for $F(x_i)$. These are usually employed to accommodate an assumed underlying distributional form such as a normal, log-normal, Weibull or other model (Gumbel, 1958; Wilk and Gnanadesikan, 1968). Such an assumption is not necessary for nonparametric bootstrap resampling; therefore the Equation [6.2] is perfectly adequate for our purpose.

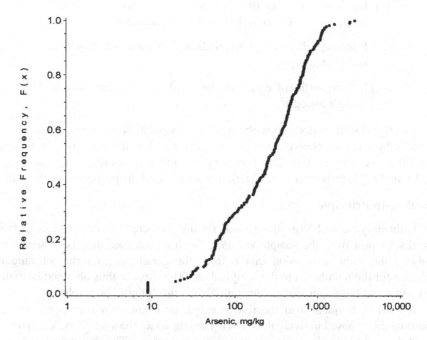

Figure 6.4 Empirical Cumulative Distribution Function, Residential Risk-Based Sampling Site #3, Vasquez Boulevard and I-70 Site, Denver, CO

This representation of data is not new. Galton by 1875 borrowed a term from architecture and called the ecdf an "ogive" (Stigler, 1986, pp. 267–72). Galton's naming convention apparently did not catch on and has not appeared in many texts on elementary statistical published since the 1960s.

The utility of the ecdf, however, was nicely described by Wilk and Gnanadesikan (1968) as follows:

> The use of the e.c.d.f. does not depend on any assumption of a parametric distributional specification. It may usefully describe data even when random sampling is not involved. Furthermore, the e.c.d.f. has additional advantages, including:
>
> (i) It is invariant under monotone transformation, in the sense of quantiles (but not, of course, appearance).
>
> (ii) It lends itself to graphical representation.
>
> (iii) The complexity of the graph is essentially independent of the number of observations.
>
> (iv) It can be used directly and valuably in connection with censored samples.

(v) It is a robust carrier of information on location, spread, and shape, and an effective indicator of peculiarities.

(vi) It lends itself very well to condensation and to interpolation and smoothing.

(vii) It does not involve the "grouping" difficulties that arise in using a histogram.[*]

Clearly all of the information about the true population cumulative distribution provided by a representative sample is also contained in the ecdf for that sample. Therefore, the ecdf is a *sufficient statistic* for the true distribution. Efron and Tibshirani (1993) provide a more detailed description of the properties of the ecdf.

The Plug-In Principle

Ultimately we wish to estimate an interesting characteristic, or *parameter*, of the true distribution from the sample at hand. Such an estimate may be obtained by applying the same expression that defines the parameter in terms of random variables for the population to the sample data. The *statistic* thus obtained becomes the estimate of the population parameter. This is the "plug-in principle."

Interest is frequently in the true average, or mean, exposure concentration assuming the exposed individual moves at random about the site. This is frequently called the *expected* value of the exposure concentration. The theoretical expected value of exposure concentration, $E(x)$, is given by

$$E(x) = \int_0^x x\left(\frac{dF^*}{dx}\right) dx \qquad [6.3]$$

Here F^* represents the true cumulative distribution.

The plug-in estimate of $E(x)$ is precisely the same function applied to the ecdf. Because the sample is finite, we replace the integration with a summation.

$$\hat{E}(x) = \sum_{i=1}^{N} x_i[F(x_i) - F(x_{i-1})] = \sum_{i=1}^{N} \left(x_i \frac{1}{N}\right),$$

or   [6.4]

$$\bar{x} = \frac{\sum_{i=1}^{N} x_i}{N}$$

which is the sample mean or arithmetic average.

---

[*] Wilk, M. B. and R. Gnanadesikan, "Probability Plotting Methods for the Analysis of Data," *Biometrika*, 1968, 55, 1, pp. 1–17, by permission of Oxford University Press.

The Bootstrap

The application of bootstrap resampling is quite simple. One repeatedly and randomly draws, with replacement, samples from the original representative sample and calculates the desired statistic. The distribution of the statistic thus repeatedly calculated is the distribution of the estimates of the sought-after population parameter. The quantiles of this distribution can then be used to establish "confidence" limits for the desired parameter.

The applicability of "random" sampling deserves some additional comment. The classic text by Dixon and Massey (1957) defines random sampling as follows:

> **Random Sampling.** When every individual in the population has an equal and independent chance of being chosen for a sample, the sample is called a *random sample*. Technically every individual chosen should be measured and returned to the population before another selection is made ....

> It might well be pointed out that saying, "Every individual in the population has an equal chance of being in the sample" is not the same as saying, "Every measurement in the universe has an equal chance of being in the sample."

By using the fact that the distribution of the ecdf is uniform between zero and one, random samples may easily be drawn using any good generator of random numbers between zero and one.

Before continuing with the full Site #3 example, it is instructive to illustrate bootstrap resampling with an example that will fit on a page. Consider the following arsenic concentrations, x_i, arising from 10 surface soil samples. The sample results have been ordered according to magnitude to facilitate the calculation of the ecdf using Equation [6.2].

Table 6.1
Original Sample

x_i	9	43	65	107	183	257	375	472	653	887
$F(x_{i)}$	0.1	0.2	0.3	0.4	0.5	0.6	0.7	0.8	0.9	1.0
Prob	0.1	0.1	0.1	0.1	0.1	0.1	0.1	0.1	0.1	0.1

Imagine that each of the 10 observed concentration values is written on a ping-pong ball and placed into a device commonly used to draw numbers for the nightly state lottery. There are 10 balls in our lottery machine and each one has an equal chance, 0.1, of being drawn. The first ball is drawn and we see that it has a concentration of 472 mg/kg on it. This concentration value is recorded and the ball returned to the machine. A second ball is drawn. Again, as with the first draw each of the 10 balls has an equal probability of being selected. This time the concentration value on the ball is 9 mg/kg. This ball is also returned prior to making a third draw. The process is repeated until we have made 10 draws. The collection of the 10 resulting concentration values constitutes the first bootstrap resample. This sample is given as the first line in Table 6.2.

The concentration values given in Table 6.2 are listed left to right in the order in which they were drawn. Note that concentrations of 9 mg/kg and 887 mg/kg appear twice in the first bootstrap resample. This is a consequence of the random selection. Assuming that the original sample is representative of possible exposure at the site and exposure will occur at random then the first bootstrap resample provides as good information on exposure as the original sample.

Table 6.2
Bootstrap Resampling Example

Bootstrap Sample	Resampled Arsenic Concentrations, mg/kg										Mean
1	472	9	887	257	9	887	375	43	183	653	377.5
2	183	183	9	653	887	9	257	472	653	257	356.3
3	43	43	472	183	257	887	257	887	9	472	351
4	43	653	653	183	653	107	43	257	472	375	343.9
5	65	257	9	887	653	257	472	9	183	887	367.9
.	.	.	.	.	.	.	.	.	.	.	.
.	.	.	.	.	.	.	.	.	.	.	.
.	.	.	.	.	.	.	.	.	.	.	.
4998	107	257	65	887	472	887	257	9	375	887	420.3
4999	375	375	183	887	472	9	43	257	887	65	355.3
5000	183	9	887	375	887	257	887	43	375	43	394.6
5001	472	65	107	257	257	472	375	653	183	107	294.8
5002	107	375	257	183	472	653	65	65	43	65	228.5
.	.	.	.	.	.	.	.	.	.	.	.
.	.	.	.	.	.	.	.	.	.	.	.
.	.	.	.	.	.	.	.	.	.	.	.
9996	472	375	107	9	9	65	107	183	43	107	147.7
9997	887	257	375	375	43	65	9	107	183	183	248.4
9998	183	43	107	9	653	107	107	887	65	887	304.8
9999	653	375	107	43	653	107	887	375	43	65	330.8
10000	107	183	9	653	887	9	65	472	375	107	286.7

The process of selecting a set of 10 balls is repeated a large number of times, say, 5,000 to 10,000 times. The statistic of interest, the average exposure concentration, is calculated for each of the 5,000 to 10,000 bootstrap resamples. The resulting distribution of the statistics calculated from each of the bootstrap resamples provides a reasonable representation of the population of all such estimates.

It is important that each bootstrap resample be of the same size as the original sample so that the total amount of information, a function of sample size, is preserved.

The validity of "random sampling" assumption is not always immediately evident. We will revisit the example Site #3 again in Chapter 7 when we investigate the spatial correlation among these 224 samples. There it will be demonstrated that arsenic is not distributed at random over this residential property, but that the arsenic concentrations are spatially related. **It is only with the assumption that the exposed individual moves around the site at random that estimates, via the bootstrap or classical methods, can be considered as reasonable.** If this assumption cannot be embraced, then other more sophisticated statistical techniques (Ginevan and Splitstone, 1997) must be employed to make reasonable decisions.

To review, the only assumptions made to this point are:

- The sample truly represents the population of interest; and,
- The principle of random sampling applies.

These assumptions must be acknowledged at some point in employing *any* statistical procedure. There are no additional assumptions required for the bootstrap.

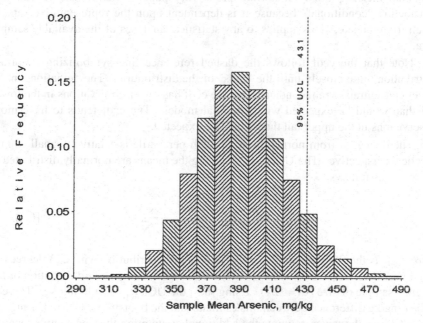

**Figure 6.5 10,000 Bootstrap Means — Sample Size 224,
Residential Risk-Based Sampling Site #3,
Vasquez Boulevard and I-70 Site, Denver, CO**

Bootstrap Estimation of the 95% UCL

The most straightforward bootstrap estimator of the 95% UCL is simply the 95th percentile of the bootstrap distribution of plug-in estimates of the sample mean arsenic concentration. The distribution of 10,000 plug-in estimates of the mean of samples of size 224 is shown in Figure 6.5. The 95th percentile is 431 micrograms per kilogram (mg/kg) of soil.

Generation of this distribution took less than 30 seconds using SAS on a 333-mhz Pentium II machine with 192 megabytes of random access memory.

Application of the Central Limit Theorem

This distribution appears to support the applicability of the Central Limit Theorem. This powerful theorem of statistics asserts that as the sample size becomes large, the statistical distribution of the sample mean approaches the "normal" or Gaussian model. A sample size of 224 is "large" for environmental investigations, particularly for parcels the size of a residential lot. However, applicable tests (see Chapter 2) indicate a statistically significant departure from normality. Many elementary statistics texts and USEPA guidance incorrectly suggest that normality of the distribution of the sample mean is a viable assumption when a sample size of 30 has been achieved.

Figure 6.6 presents conditional cumulative distribution function (ccdf) of bootstrap means plotted on normal probability paper. We choose to call this distribution "conditional" because it is dependent upon the representative sample taken from the site. This applies to any statistical analyses of the available sample data.

Note that the ccdf follows the dashed reference line symbolizing a normal distribution quite closely until the "tails" of the distribution. The deviations in the lower concentration range indicate that the ccdf has fewer observations in the lower tail than would be expected with a normal model. The ccdf tends to have more observations in the upper tail than would be expected.

The deviation from normality at the 95th percentile is relatively small from a practical perspective. The UCL_{norm} assuming the means are normally distributed is estimated as follows:

$$UCL_{norm} = \bar{x} + t_{0.95}\frac{s}{\sqrt{N}} \qquad [6.5]$$

Here, $t_{0.95}$ is the 95th percentile of the student "t" distribution with 223 degrees of freedom, 1.645. $\bar{X}$ (= 385) and s (= 407) are the mean and standard deviation of the original sample of size N (= 224). Using [6.5] the UCL_{norm} is 430 mg/kg. There is only 1 mg/kg difference between this estimate and the bootstrap UCL of 431 mg/kg. The UCL_{norm}, however, requires the additional assumption that the sample mean is normally distributed. This assumption may be in doubt for extremely skewed distributions of the original data and smaller sample sizes.

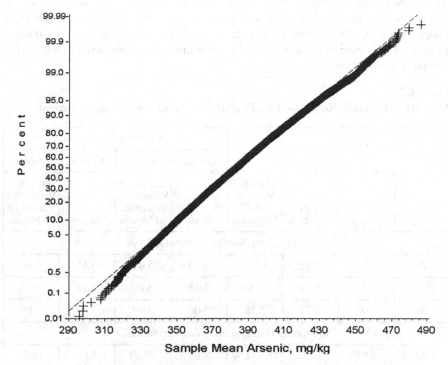

Figure 6.6. CCDF 10,000 Bootstrap Means — Sample Size 224, Residential Risk-Based Sampling Site #3, Vasquez Boulevard and I-70 Site, Denver, CO

The Bootstrap and the Log-Normal Model

The log-normal density model is often used to approximate the distribution of environmental contaminants when their distribution is right skewed. Before the ready availability of convenient computing equipment, Land (1975) developed a method, and provided tables, that can be used to estimate a UCL on the arithmetic mean of a log-normal distribution. If we take the mean, $\bar{y}$, and standard deviation, s_y, of the log-transformed data then a $(1-\alpha)$ UCL for the arithmetic mean is defined by:

$$UCL = e^{\left[\bar{y} + 0.5s_y^2 + s_y H_{1-\alpha}(n-1)^{\frac{1}{2}}\right]} \qquad [6.6]$$

where n is the sample size and $H_{1\alpha}$ is a tabled constant that depends on n and s_y. Tables of $H_{1-\alpha}$ are given by Land (1975) and Gilbert (1987). This method will provide accurate UCLs on the arithmetic mean if the data are truly distributed as a log-normal. The Land procedure is recommended in EPA guidance (USEPA, 1989).

In order to investigate the relative merits of Land's estimator and the bootstrap estimator when the assumption of log-normality is not satisfied, the authors conducted a Monte Carlo simulation (Ginevan and Splitstone, 2002). This study considered estimator performance for both a true log-normal model and a mixture of four log-normal models. Each distributional case was investigated for three typical sample sizes of 20, 40, and 80 samples. In addition, the effect of approximate 10 percent censoring in the lower concentration range was considered as well as the complete sample case.

Table 6.3
Monte Carlo Study Result — Land's Estimator and Bootstrap Estimator

	Simulated Distribution			Bootstrap Estimators			
				Complete Samples		Censored Samples	
Sample Size	Mean	95th Percentile	Expected Value of Land's Estimator	Mean	Expected Value 95th Percentile of Means	Mean	Expected Value 95th Percentile of Means
Single Log-Normal Distribution							
20	3.10	6.33	13.84	3.10	5.46	3.11	5.46
40	3.09	5.34	7.03	3.13	5.00	3.45	5.00
80	3.08	4.75	5.15	3.08	4.48	3.43	4.09
Mixture of Four Log-Normal Distributions							
20	8.67	22.13	18,734.98	8.68	19.03	8.68	19.03
40	8.45	18.90	622.04	8.45	16.41	8.46	16.41
80	8.50	16.73	164.66	8.50	14.88	8.50	14.88

It has been the author's experience that while most environmental data appear to follow a distribution that is skewed to the right, rarely does it follow a pure log-normal model. Further, most site investigations include observations reported as below the limit of "method detection" or "quantification." Values were estimated for the censored sample results as the expectations using regression on expected normal scores as recommended by Helsel (1990) or the method of maximum likelihood (Millard, 1997). The results of this Monte Carlo study are briefly summarized in Table 6.3.

Perhaps the most striking finding of this study is the extremely poor performance of the Land estimator. These results are actually to be expected. The Land estimator is extremely dependent on the value of the sample standard deviation, s_y, because it plays a major role in Equation [6.6] and also is a major determinant of the tabulated value of H.

One might complain that the bootstrap UCL is biased slightly low relative to the Monte Carlo bound from the 5,000 samples drawn from the parent distribution. This is clearly so, and is to be expected given the extreme right skew of the parent distribution. However, this bias is only apparent because we "know" the "true" distribution. In practice, while we know that the population distribution has a long tail, we do not know what the exact form of this tail is. Assuming that the ecdf adequately characterizes the tail of the parent distribution appears to have far less consequence than making an incorrect assumption regarding the form of the parent distribution.

Pivotal Quantities

Singh, Singh, and Engelhardt (1997) investigate bootstrap confidence limits for the mean of a log-normal distribution in their issue paper prepared for the USEPA. They suggest that these confidence limits should be estimated via "pivotal" quantities. One such estimate uses the mean, $\bar{x}_B$, of the bootstrap ccdf of sample means and its standard error, $s_{\bar{x}_B}$, in the following relationship:

$$UCL = \bar{x}_B + z_{0.95} s_{\bar{x}_B} \qquad [6.6]$$

Here $Z_{0.95}$ is the 95th percentile of the standard normal distribution. This estimator requires the assumption that the distribution of the bootstrap sample means is normally distributed. This is a debatable assumption even for reasonably large samples as indicated above and appears to be an unnecessary complication.

Singh, et al. also consider the "bootstrap t," which is simply formed for each bootstrap sample by subtracting the mean of the original sample, $\bar{x}$, from the mean of the ith bootstrap sample, $\bar{x}_i$, and dividing the result by the standard deviation of the bootstrap sample, $s_{x,i}$. The distribution of the pivotal t,

$$t_i = \frac{\bar{x}_i - \bar{x}}{s_{x,i}} \qquad [6.7]$$

is then found and the desired percentile of the pivotal t distribution is applied to the mean and standard deviation of the original sample to obtained the desired UCL estimate. Efron and Tibshirani (1993, p. 160) clearly indicate that this method gives erratic results and its use appears to be an unnecessary complication to the estimation of the UCL.

Bootstrap Estimation of CCDF Quantiles

Environmental interest is not always focused on the average, or mean, of the underlying data distribution. One of the oldest environmental regulations necessitates the determination of the 99th percentile of daily wastewater effluent concentration. This percentile defines the "daily limit" applicable to facilities regulated under the National Pollution Discharge Elimination System (NPDES) mandated by the Clean Water Act.

The U. S. Environmental Protection Agency (USEPA) recognized the statistical realities associated with the control and measurement of water discharge parameters as early as 1974 in the effluent guidelines and standards applicable to the Iron and Steel Industry (Federal Register Vol. 39, No. 126, p. 24118, Friday, June 28, 1974). Paragraph (6) of (b) *Revision of proposed regulations prior to promulgation* clearly suggests that there was early concern that established limitations may be frequently exceeded by effluent of a "well designed and well operated plant."

> (6) As a precaution against the daily maximum limitations being violated on an intolerably frequent basis, the daily maximum limitations have been increased to three times the values permitted on the "30 consecutive day" basis .... The daily limits allow for normal daily fluctuations in a well design and well operated plant ....

The largely ad hoc "three times" criterion has been supplanted by more statistically sophisticated techniques over time. This evolutionary regulatory process is nicely summarized by Kahn and Rubin (1989):

> An important component of the process used by EPA for developing limitations is the use of entities referred to as variability factors. These factors are ratios of high effluent to average levels that had their origin as engineering "rules of thumb" that express the relationship between average treatment performance levels and large values that a well designed and operated treatment system should be capable of achieving all the time. Such factors are useful in situations where little data are available to characterize the long-term performance of a plant or group of plants. As the effluent guidelines regulatory program evolved, the development of these variability factors became more formalized, as did many other aspects of the program, in response to legal requirements to document thoroughly ....[*]

As a result of this evolutionary regulatory program, the daily maximum limitation is generally considered an estimate of the 99th percentile of the statistical distribution of possible daily effluent measurement outcomes of a "well designed and well operated plant." The monthly average or "30-day" limitations are generally considered to be the 95th percentile of the statistical distribution of possible "30-day" average effluent measurement outcomes of a "well designed and well operated plant." Thus, the effluent of a "well designed and well operated plant" is *expected* to exceed the effluent limitation one percent of the time for daily measurements and five percent of the time for "30-day" average values. (See Kahn and Rubin, 1989; Kahn, 1989; USEPA, 1985 App. E; USEPA, 1987; USEPA, 1993).

The use of percentiles of the statistical distribution of measurement outcomes as permit limitations has been widely, and consistently, publicized by the USEPA. Indeed, it has been discussed in versions of the USEPA's *Training Manual for NPDES Permit Writers* issued five years apart (USEPA, 1987; USEPA, 1993).

---

[*] Kahn, H. and Rubin, M., "Use of Statistical Methods in Industrial Water Pollution Control Regulations in the United States," *Environmental Monitoring and Assessment*, 12: 129–148, 1989, Kluwer Academic Publishers. With permission.

> Regulatory agencies have settled on a statistical confidence rate of
> 1% to 5% (typically, 1% rates for daily maximum, 5% rate for
> monthly average). These confidence rates correspond to the 99th to
> 95th percentiles of a cumulative probability distribution .... Thus,
> a discharger running a properly operated and maintained treatment
> facility has a 95–99% chance of complying with its permit limits in
> any single monitoring observation. (USEPA, 1987, p. 17)

> When developing a BPJ [Best Professional Judgment] limit,
> regulatory agencies have settled on a statistical confidence rate of
> 1 to 5. These confidence rates correspond to the 99th to 95th
> percentiles of a cumulative probability distribution .... Thus, in
> any single monitoring observation, a discharger running a properly
> operated and maintained treatment facility has a 95–99% chance
> of complying with its permit limits. (USEPA, 1993, pp. 3–5)

> Determining effluent limitations is roughly analogous to
> establishing industrial quality control limits in that process data
> are used in a statistical analysis to determine bounds on measures
> that indicate how well the process is being operated. (Kahn and
> Rubin, 1989, p. 38)

Clearly, these statements suggest that the USEPA assumes that the variation in waste water discharge concentrations from a "well designed and well operated plant" is random. Unfortunately, The USEPA also relies heavily on the assumption that the statistical distribution of effluent concentration follows a log-normal model (Kahn and Rubin, 1989). This assumption is, in many cases, unjustified.

While the "formalized" effluent guidelines regulatory program employs the language of statistics, it remains largely based upon the use of "best engineering judgment."

Mega-Hertz Motor Windings, Inc. discharges treatment plant effluent to the beautiful Passaic River. Their treatment plant is recognized as an exemplary facility, however, the effluent concentration of copper does not meet the discharge limitation for their industrial subcategory. Mega-Hertz must negotiate an NPDES permit limitation with their state's Department of Environmental Protection.

The historical distribution of measured daily copper concentrations is shown in Figure 6.7. Note that this distribution of 265 daily values is skewed to the right but differs significantly from log-normality. The Shapiro-Wilk test indicates that there is only a 0.0002 probability of the data arising from a log-normal model. Figure 6.8 clearly shows that there are significant deviations from log-normality in the tails of the distribution.

Assuming a log-normal model will overestimate the 99th percentile of concentration used for the daily discharge limitation. However, ASSUMING that the collected data are truly representative of discharge performance and that sampling will be done on random days, then a bootstrap estimate of the 99th percentile will nicely serve the need. There is no need to assume any particular distributional model.

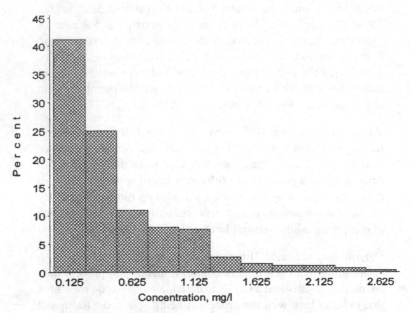

**Figure 6.7   Daily Discharge Copper Concentrations,
Mega-Hertz Motor Winding Treatment Plant**

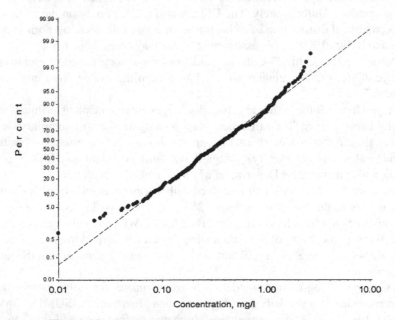

**Figure 6.8   Log-Normal Probability Plot,
Daily Discharge Copper Concentrations,
Mega-Hertz Motor Winding Treatment Plant**

Bootstrap Quantile Estimation

The bootstrap estimation of a specific quantile of the population distribution is in concept no different than estimating the mean. The ecdf is repeatedly and randomly sampled with replacement and the plug-in principle is used to estimate the desired quantile. In the current example, the 99th percentile of concentration is given by the 263 highest observation in the bootstrap sample of size 265. In other words, the third ranked from highest observation in the sample is used as the estimate of the 99th percentile.

The general method for determining which of the ranked observations to use as the estimate of the 100Pth percentile in a sample of size N is as follows:

Order the sample from lowest to highest as indicated in relation [6.1] above. Let

$$NP - j + g \qquad\qquad [6.9]$$

where j is the integer part and g is the fractional part of NP. If g = 0 then the estimate is observation X_j if g = 0 or X_{j+1} if g > 0. This provides an estimate based upon the "empirical distribution function" (SAS, 1990, p. 626).

Expected Value or Tolerance Limit

To estimate the daily discharge limitation for Mega-Hertz Motor Windings, Inc. NPDES permit 5,000 bootstrap samples were generated and the 99th percentile of copper concentration estimated for each bootstrap sample. Generation of these 5,000 bootstrap estimates took approximately 23 minutes using SAS on a 333-mhz Pentium II with 192 megabytes of random access memory. The large amount of time required is due directly to the need to sort each of the 5,000 bootstrap samples of size 265.

The histogram of the 5,000 bootstrap estimates of the 99th percentile of effluent copper concentration is given in Figure 6.9. Note that this distribution is left skewed. The expected value of the 99th percentile is the mean of this distribution, 2.27 milligrams per liter (mg/l). The median is 2.29 mg/l and the 95th percentile of the distribution is 2.62 mg/l. While these statistics are interesting in themselves, they beg the question as to which estimate to use as the permit discharge limit.

To help make that decision we may look at the bootstrap estimates as estimates of a "tolerance limit" (see Guttman, 1970). In the present context a tolerance limit provides an upper bound for a specified proportion (99 percent) of the daily discharge concentration of copper with a specified degree of confidence. For instance we are 50 percent confident that the median, 2.29 mg/l, provides an upper bound for 99 percent of the daily discharge concentrations.

In other words we flip a coin to determine whether the "true" 99th percentile is above or below 2.29.

Because the expected value, 2.27 mg/l, is below the median, we are less than 50 percent confident that the "true" 99th percentile is 2.27 mg/l or less. The penalties for violating NPDES permit conditions can range from $10,000 to $25,000 per day and possible jail terms. It therefore seems prudent to be highly confident that the estimate selected will indeed contain at least the desired proportion of the discharge concentrations.

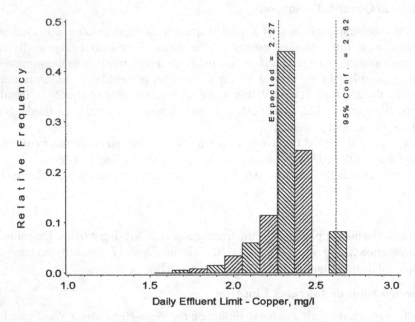

Figure 6.9   **Bootstrap Estimates of 99th Percentile,
Daily Discharge Copper Concentrations,
Mega-Hertz Motor Winding Treatment Plant**

By selecting the 95th percentile of our bootstrap distribution of estimates, we
are 95 percent confident that at least 99 percent of the daily discharges will be less
than 2.62 mg/l. Turning this around, we are 95 percent confident that there is no
more than a one-percent chance of an unintentional permit violation if the permit
limit for effluent copper is 2.62 mg/l.

We note that application of the bootstrap to the estimation of extreme
percentiles is not as robust as its application to the estimation of summary statistics
like the mean and variance (Efron and Tibshirani, 1993, pp. 81–83). It requires large
original sample sizes and the results are more sensitive to outlying observations.
Thus one should approach problems like this example with caution. However, one
question that must also be considered is "What can I do that would be any better?"
In this case a better alternative is not obvious, so, despite its limitations, the bootstrap
provides a reasonable solution.

Estimation of Uranium-Radium Ratio

The decommissioning plan for the Glo-in-Dark Products Inc. site identifies
criteria for unrestricted site release for natural uranium, and Ra_{226} in soil as follows:

- Natural Uranium — 10 pico Curies per gram (pCi/g)
  (Assumes all daughters in equilibrium, including Ra_{226} and Th_{230})

- Ra_{226} — 5 pCi/g

These criteria are based on the assumption that Ra_{226} is in equilibrium with uranium (U_{238} and U_{234}) and Th_{230}.

The method selected for analysis of soil samples at the facility, sodium iodide (NaI) gamma spectroscopy, is not capable of measuring uranium. Therefore, it is necessary to infer uranium concentrations via measurements of Ra_{226} by application of a site-specific ratio of uranium to Ra_{226}. If the ratio exceeds one, the excess uranium and thorium concentrations must be accounted for using isotope specific criteria. This ratio is incorporated into a site-specific "unity rule" calculation to apply to soil sample analysis results to determine compliance with site release criteria as described in the decommissioning plan.

A total of 62 reference samples are available for use in estimating the desired U_{238} to Ra_{226} ratio. These samples were specifically selected for use in estimating this ratio. It is assumed that the selected samples are *representative* of the U_{238} and Ra_{226} found at the site. A scatter diagram of the U_{238} versus Ra_{226} determined by gamma spectroscopy at a radiological sciences laboratory is shown in Figure 6.9.

Candidate Ratio Estimators

There are two candidate estimators of the ratio of U_{238} to Ra_{226} assuming a linear relationship that must pass through the concentration origin (0,0). Both of these estimators are derived as the best linear unbiased estimate of the slope, B, in the following relationship (Cochran, 1963, pp. 166–167):

$$y_i = Bx_i + error_i \qquad [6.10]$$

Here y_i represents the concentration of U_{238} observed at a fixed concentration of Ra_{226}, x_i. The $error_i$'s are assumed to be independent of the concentration of Ra_{226}, x_i, and have mean zero and a variance proportional to a function of x_i.

The assumption of independence between the $error_i$'s and the x_i is tenuous in practice. Certainly, both the observed concentration of U_{238} and that of Ra_{226} are subject to measurement, sampling, and media variations. We may assume that the impact of the "error" variation associated with the x_i on the $error_i$ is small when compared to that of the y_i. If this is not the case the estimate of B will be biased. While this bias is important when seeking an estimate of the theoretical ratio, it has little impact on the estimation of the U_{238} concentration from an observed Ra_{226} concentration.

The following formula for the weighted least-squares estimator, b, is used for both candidate estimators of the ratio B:

$$b = \frac{\sum w_i y_i x_i}{\sum w_i x_i^2} \qquad [6.11]$$

The difference between the estimators depends upon the assumptions made regarding the "error" (residual) variance. If the error variance is proportional to the Ra_{226} concentration then the "best" estimate uses the weight,

$$w_i = \frac{1}{x_i}$$

With this weight the "best" least-squares estimate, b, is simply the ratio of the mean U_{238} concentration, $\bar{y}$, to the mean Ra_{226} concentration, $\bar{x}$. In other words,

$$b = \bar{y}/\bar{x} \qquad\qquad [6.12]$$

Sometimes the relationship between y_i and x_i is a straight line through the origin, but the error variance is not proportional to x_i but increases roughly as x_i^2. In this case the "best" estimate uses the weight,

$$w_i \propto \frac{1}{x_i^2}$$

The least-squares estimate of b is then the mean ratio over the sampling units:

$$b = \frac{\sum w_i y_i x_i}{\sum w_i x_i^2} = \frac{1}{n}\sum\left(\frac{y_i}{x_i}\right) \qquad\qquad [6.13]$$

The data suggest that this latter estimator may be the most appropriate as discussed below.

Note that the usual unweighted linear least-squares estimator, i.e., $w_i = 1$, is not appropriate because it requires a constant error variance over the range of data. This is obviously not the case.

Data Evaluation

Figure 6.10 gives a scatter diagram of the concentration of U_{238} versus the concentration of Ra_{226} as measured in the 62 reference sample, for this site. Note that most of the samples exhibit low concentrations, less than 100 pCi/gm, of U_{238}. This suggests that the U_{238} and Ra_{226} concentrations at this site are heterogeneous and might result from a mixture of "statistical" bivariate populations of U_{238} and Ra_{226} concentrations. This suspicion is further confirmed by looking at the empirical frequency distribution of individual sample ratios of U_{238} to Ra_{226} concentration given as Figure 6.11. This certainly suggests that the site-specific distribution of the U_{238} to Ra_{226} ratios is highly skewed with only a few ratios greater than 2.0.

The data scatter exhibited in Figure 6.10 also suggests that the "error" variation U_{238} concentration is not linearly related to the concentration of Ra_{226}. This favors the "mean ratio" estimator [6.13] as the site-specific ratio estimator of the mean U_{238} to Ra_{226} ratio. Because the empirical distribution of the U_{238} to Ra_{226} ratios from the reference samples is so skewed, it is unlikely that the mean ratio from even 62 samples will conform to the Central Limit Theorem and approach a normal or Gaussian distribution. Therefore, the bootstrap distribution of the mean ratio estimate was constructed from 10,000 resamplings of the reference sample data with replacement.

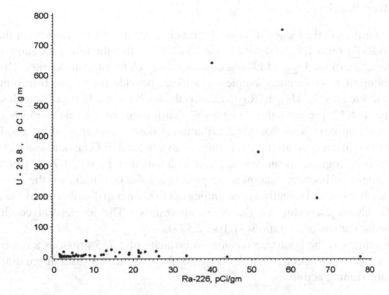

Figure 6.10   U-238 versus Ra-226 Concentrations;
Soil Samples — Glo-in-Dark Products Site

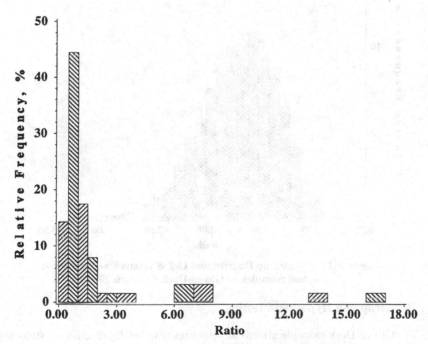

Figure 6.11   U-238 versus Ra-226 Concentration Ratios;
Soil Samples — Glo-in-Dark Products Site

Bootstrap Results

Regardless of the techniques used for making statistical inferences regarding the U_{238} to Ra_{226} ratio at this site, the *key assumption* is that the reference samples are representative of the U_{238} and Ra_{226} concentration relationship at the site. Thus the resampling of the reference samples will likely provide for as much information about the site specific U_{238} to Ra_{226} ratio as the analysis of additional samples.

Figure 6.12 presents the "bootstrap" distribution of the mean ratios of 62 individual samples. Note that this distribution is skewed and appropriate goodness-of-fit tests indicate a significant departure from a normal or Gaussian distribution.

The bootstrap site mean ratio is 1.735 with a median ratio of 1.710. Due to the large number of bootstrap samples, 95 percent confidence limits for the site mean ratio may be obtained directly as percentiles of the bootstrap distribution as the 250th and 9750th largest values of the bootstrap results. The 95 percent confidence interval for the site mean ratio is (1.106, 2.524).

In summary, the bootstrap mean ratio estimate of 1.735 provides a reasonable estimate of a site-specific single U_{238} to Ra_{226} ratio for use in guiding site decommissioning activities.

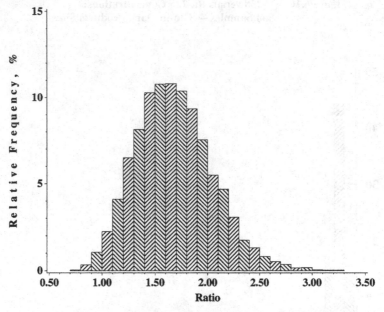

Figure 6.12   Bootstrap Distribution U-238 versus Ra-226 Ratios;
Soil Samples — Glo-in-Dark Products Site

The Bootstrap and Hypothesis Testing

The Glo-in-Dark example affords an opportunity to briefly discuss the statistical testing of hypothesis as supported by the bootstrap. Mary Natrella's (1960) nice article on the relationship between confidence intervals and tests of significance provides the link between the two. In the present example suppose that we wish to test the hypothesis that the true mean ratio is one (1.0). Clearly, one lies outside of the bootstrap 95 percent confidence interval about the true mean ratio (1.106, 2.524). Therefore, we may conclude that the true mean ratio is significantly different from one at the 0.05 level of significance.

It had been previously hypothesized that the mean U_{238} to Ra_{226} ratio was 1.7. Clearly, 1.7 lies near the center of the 95 percent confidence interval about the true mean ratio. Therefore, there is no evidence to suggest that the mean ratio is significantly different from 1.7 at the 0.05 level of significance. Comparison of the bootstrap distributions of the same statistic derived from representative samples from two different populations is also entirely possible using quantile-quantile plot techniques described by Wilk and Gnanadesikan (1968).

The Bootstrap Alternative to the Two-Sample t-test

Sometimes we encounter data sets that are not well suited to either parametric or rank-based nonparametric methods. Consider the data set presented in Table 6.4. These represent two hypothetical samples. Each has 30 observations.

The first sample (x1) is a mixture of 10 random observations each from log-normal distributions with geometric means of 0.1, 1.0, and 10.0 and a geometric standard deviation of 4.5 in each population. The second sample (x2) is from a "pure" log normal with a geometric mean of 1 and a geometric standard deviation of 4.5. The last five lines of Table 6.4 present arithmetic means, standard deviations (SD), minima, maxima, and medians for the 2 samples and for the natural logarithms of the observations in the two samples.

One might ask, "Isn't this a terribly contrived data set?" The answer is: not really. Let us suppose that we had two areas, one that was known to be uncontaminated (thus representing "background") and the other, which was thought to be contaminated. Let us suppose further that roughly 1/3 of the contaminated area was unaffected in the sense that it was already at background and that another 1/3 was remediated in a way that the existing soil was replaced by clean fill, that had lower levels of the material of interest than the background soils in the area. The problem is of course that 1/3 was contaminated, but not remediated. Thus we have 1/3 very clean, 1/3 background, and 1/3 contaminated. The question is, "Did the remediation remove enough contaminated material to make the two areas equivalent from a risk perspective?" Since arithmetic means are usually the best measure of average contamination for the purposes of risk assessment, the statistical question is, "Do the two areas have equivalent arithmetic mean concentrations?"

The first observation is that the arithmetic means appear very different, so one would be inclined to reject the idea that the two areas are equivalent. However, the second observation is that the standard deviations and thus variances are also very different, so we do not want to employ a t-test. We might try doing a logarithmic transformation of the data. Here there is still some difference in means and standard deviations, but much less so than the original observations. Unfortunately, a t-test done on the log-transformed data, gives a p-value of about 0.06, which does not really support the idea that the two samples are different. We might also try a rank

Table 6.4
A Sample Data Set Where Neither t-Tests nor Rank Sum Tests Work Well

	Mixture Distribution X1	Single Distribution X2	Ln(X1)	Ln(X2)
	0.26	0.62	−1.3425	−0.4713
	0.06	0.54	−2.7352	−0.6105
	0.05	0.95	−3.0798	−0.0486
	0.61	2.42	−0.4993	0.8840
	0.08	0.51	−2.5531	−0.6724
	0.08	0.52	−2.4963	−0.6513
	0.03	0.41	−3.6467	−0.8884
	0.02	0.81	−3.8882	−0.2158
	0.11	0.01	−2.1680	−4.3020
	0.11	0.43	−2.2486	−0.8456
	0.31	4.40	−1.1565	1.4812
	7.32	0.31	1.9901	−1.1638
	0.61	0.12	−0.4897	−2.1112
	0.22	2.02	−1.5209	0.7051
	0.49	2.25	−0.7040	0.8097
	0.48	0.43	−0.7398	−0.8452
	0.58	3.32	−0.5440	1.1996
	10.12	1.83	2.3143	0.6069
	2.19	0.26	0.7857	−1.3363
	10.54	0.69	2.3554	−0.3772
	1.34	0.05	0.2924	−3.0053
	2.49	2.08	0.9133	0.7306
	17.51	1.16	2.8626	0.1455
	32.18	4.17	3.4713	1.4273
	25.77	0.27	3.2493	−1.3031
	262.87	0.25	5.5717	−1.3856
	3.76	0.12	1.3244	−2.1452
	31.39	3.66	3.4465	1.2963
	147.69	4.83	4.9951	1.5738
	45.12	0.24	3.8093	−1.4455
Mean	20.147	1.322	0.2523	−0.4321
SD	53.839	1.440	2.6562	1.3947
Minimum	0.021	0.014	−3.8882	−4.3020
Maximum	262.870	4.8251	5.5717	1.5738
Median	0.6099	0.5836	−0.4945	−0.5409

sum test. However, the central tendency in terms of the sample medians is virtually the same, so this test is not even remotely significant. Thus we are left with an uncomfortable suspicion that the arithmetic means are different but with no good way of testing this suspicion.

Bootstrap to the Rescue!

If we had just one sample, we know that we could construct confidence bounds for the mean by simply resampling the sample data. It turns out that we can get something that looks much like a standard t-test in the same way. The process involves three steps:

1. For x1 and x2 we can resample 10,000 times and come up with two sets of bootstrap sample means, $\bar{x}1b_i$ and $\bar{x}2b_i$.

2. We generate $\bar{x}1b_i$ and $\bar{x}2b_i$ using a random number generator, thus both sets of means are generated in random order. Thus, the difference $\bar{x}1b_i - \bar{x}2b_i$, which we will refer to as $\bar{x}d_i$, is a bootstrap distribution for $\mu1 - \mu2$. That is, we can simply take the differences between $\bar{x}1b_i$ and $\bar{x}2b_i$, in the order that the two sets of bootstrap samples were generated as a bootstrap distribution for $\mu1 - \mu2$.

3. Once we have obtained our set of 10,000 or so $\bar{x}d_i$ values, we can sort it from smallest to largest. If zero is less than the 251st smallest value in the resulting empirical cumulative distribution, or more than the 9750th largest value, we have evidence that the actual difference of $\mu1 - \mu2$ is significantly different from zero.

Figure 6.13 shows a histogram of 10,000 $\bar{x}d_i$ values for the samples in Table 6.4. It is evident that almost all of the differences are greater than zero. In fact only two of the 10,000 total replications have differences less than or equal to zero. Thus the bootstrap provides a satisfying means of demonstrating the difference in arithmetic means for our otherwise vexing example.

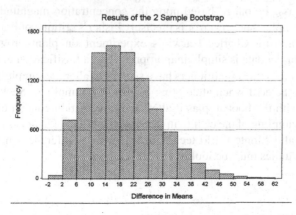

Figure 6.13 The Distribution of $\bar{x}d_i$ for the Data in Table 6.4

Epilogue

In the opinion of the authors, the bootstrap offers a viable means for assisting with environmental decision making. The only assumptions underlying the application of the bootstrap are:

- The data at hand are representative of the population of interest.

- Random sampling of the data at hand is appropriate to estimate the characteristic of interest.

These assumptions are required in justifying the application of *any* statistical inference from the sample data at hand. Many other popular statistical procedures require the additional assumption of a specific form to the underlying statistical model of random behavior. Often, these assumptions remain unjustified.

It should not be construed that the authors believe that the bootstrap offers the solution to every problem. Certainly, when statistical design permits description by well-defined linear, or nonlinear, models making the assumption of a Gaussian error model, then use of traditional parametric techniques are appropriate.

One might ask how small a sample size can be used with the bootstrap. We have found that it can work fairly well for samples as small as 10 and can be employed with some confidence with samples of 30 or more. Chernick (1999) gives a formula for the number of unique bootstrap samples, U, that can be drawn from an original sample of size N. It is:

$$ U = (2N - 1) !/[N!(N - 1)!] \qquad [6.14] $$

Here ! is the factorial notation where, for example 3! is equal to 3x2x1, or 6. For 20 observations, the number of unique samples is about 69 billion, and even for 10 samples, we have over 92,000 unique values.

It must be recognized that there are those who might suggest that the representative data giving the U_{238} and Ra_{226} found at the site be rearranged so that the U_{238} and Ra_{226} be paired based upon the concentration magnitude rather than their physical sample. Such an error is precisely Fisher's critique of Francis Galton's analysis of data from Charles Darwin's experiment on plant growth. Such a reorganization of the data is simply inappropriate and a falsification of the facts.

As a final note, some practitioners have seen fit to alter the sample size from that associated with the ecdf when attempting bootstrap resampling. Such a practice is not consistent with the bootstrap as it alters the information content of the sample. Using random sampling of the ecdf to investigate the effect of changes in sample size is certainly a valid Monte Carlo technique. However, inferences made from the results of such studies must be looked at with caution.

References

Chernick, M. R., 1999, *Bootstrap Methods: A Practitioner's Guide*, John Wiley, New York.

Cochran, W. G., 1963, *Sampling Techniques, 2nd Edition*, John Wiley & Sons, Inc., New York.

Dixon, W. J. and Massey, F. J., 1957, *Introduction to Statistical Analysis*, McGraw-Hill, New York.

Efron, B., and Tibshirani, R. J., 1998, *An Introduction to the Bootstrap*, Chapman & Hall/CRC, Boca Raton, FL.

Fisher, R. A., 1966, *The Design of Experiments, 8th ed.*, Hafner, New York.

Gilbert, R. O., 1987, *Statistical Methods for Environmental Pollution Monitoring*, Van Nostrand Reinhold, New York.

Ginevan, M. E. and Splitstone, D. E., 1997, "Risk-Based Geostatistical Analysis and Data Visualization: Improving Remediation Decisions for Hazardous Waste Sites." *Environmental Science & Technology*, 31: 92–96.

Ginevan, M. E. and Splitstone, D. E., 2002, "Bootstrap Upper Bounds for the Arithmetic Mean, and the Use of Censored Data," *Environmetrics*, 13: 1–12.

Gumbel, E. J., 1958, *Statistics of Extremes*, Columbia University Press, New York.

Guttman, I., 1970, *Statistical Tolerance Regions: Classical and Bayesian*, Hafner Publishing Co., Darien, CT.

Helsel, D. R., 1990, "Less Than Obvious: Statistical Treatment of Data below the Detection Limit," *Environmental Science and Technology*, 24: 1766–1774.

ISSI Consulting Group, 1999, "Draft Report for the Vasquez Boulevard and I-70 Site, Denver, CO.; Residential Risk-Based Sampling, Stage I Investigation," for the USEPA, Region 8, Denver, CO.

Kahn, H. and Rubin, M., 1989, "Use of Statistical Methods in Industrial Water Pollution Control Regulations in the United States," *Environmental Monitoring and Assessment*, 12: 129–148.

Kahn, H., 1989, Memorandum: "Response to Memorandum from Dr. Don Mount of December 22, 1988," U.S. EPA, Washington, D.C., to J. Taft, U.S. EPA, Permits Division, Washington, D.C., August 30, 1989.

Land, C. E., 1975, "Tables of Confidence Limits for Linear Functions of the Normal Mean and Variance," *Selected Tables in Mathematical Statistics, Vol. III*, American Mathematical Society, Providence, RI, pp. 385–419.

Millard, S. P., 1997, *Environmental Stats for S-Plus*, Probability, Statistics and Information, Seattle, WA.

Natrella, Mary G., 1960, "The Relation Between Confidence Intervals and Test of Significance," *The American Statistician*, 14: 20–23.

SAS, 1990, *SAS Procedures Guide, Version 6, Third Edition*, SAS Institute Inc., Cary, NC.

Singh, A. K., Singh, A., and Engelhardt, M., 1997, "The Lognormal Distribution in Environmental Applications," USEPA, ORD, OSWER, EPA/600/R-97/006.

Stigler, S. M., 1986, *The History of Statistics, The Measurement of Uncertainty before 1900*, The Belknap Press, Cambridge, MA.

USEPA, 1985, *Technical Support Document for Water Quality-Based Toxics Control*, NTIS, PB86-150067.

USEPA, 1987, *Training Manual for NPDES Permit Writers*, Technical Support Branch, Permits Division, Office of Water Enforcement and Permits, Washington, DC.

USEPA, 1989, *Risk Assessment Guidance for Superfund: Human Health Evaluation Manual — Part A*, Interim Final, United States Environmental Protection Agency, Office of Emergency and Remedial Response, Washington, D.C.

USEPA, March 1993, *Training Manual for NPDES Permit Writers*, EPA 833-B-93-003, NTIS PB93-217644.

Wilk, M. B. and Gnanadesikan, R., 1968, "Probability Plotting Methods for the Analysis of Data," *Biometrika*, 55(1): 1–17.

CHAPTER 7

Tools for the Analysis of Spatial Data

There is only one thing that can be considered to exhibit random behavior in making a site assessment. That arises from the assumption adopted by risk assessors that exposure is random. In the author's experience there is nothing that would support an assumption of a random distribution of elevated contaminant concentration at any site. Quite the contrary, there is usually ample evidence to logically support the presence of correlated concentrations as a function of the measurement location. This speaks contrary to the usual assumption of a "probabilistic model" underlying site measurement results. Isaaks and Srivastava (1989) capture the situation as follows:

> "In a probabilistic model, the available sample data are viewed as the result of some random process. From the outset, it should be clear that this model conflicts with reality. The processes that actually do create an ore deposit, a petroleum reservoir, or a hazardous waste site are certainly extremely complicated, and our understanding of them may be so poor that their complexity appears as random behavior to us, but this does not mean that they are random; it simply means that we are ignorant.

> Unfortunately, our ignorance does not excuse us from the difficult task of making predictions about how apparently random phenomena behave where we have not sampled them."

We can reduce our ignorance if we employ statistical techniques that seek to describe and take advantage of spatial correlation rather than ignore it as a concession to statistical theory. How this is done is best described by example. The following discusses one of those very few examples in which sufficient measurement data are available to easily investigate and describe the spatial correlation.

ABC Exotic Metals, Inc. produced a ferrocolumbium alloy from Brazilian ore in the 1960s. The particular ore used contained thorium, and slight traces of uranium, as an accessory metal. A thorium-bearing slag was a byproduct of the ore reduction process. Much of this slag has been removed from the site. However, low concentrations of thorium are present in slag mixed with surface soils remaining at this site.

The plan for decommissioning of the site-specified criteria for release of the site for unrestricted use. Release of the site for unrestricted use requires demonstration that the total thorium concentration in soil is less than 10 picocuries per gram (pCi/gm). The applicable NRC regulation also provides options for release with restrictions on future uses of the site. These allow soil with concentrations greater

163

than 10 pCi/gm to remain on the site in an engineered storage cell provided that acceptable controls to limit radiation doses to individuals in the future are implemented.

In order to facilitate evaluation of decommissioning alternatives and plan decommissioning activities for the site, it was necessary to identify the location, depth, and thickness of soil-slag areas containing total thorium, thorium 232 (Th_{232}) plus thorium 228 (Th_{228}), concentrations greater than 10 pCi/gm. Because there are several possible options for the decommissioning of this site, it is desirable to identify the location and estimated volumes of soil for a range of total thorium concentrations. These concentrations are derived from the NRC dose criteria for release for unrestricted use and restricted use alternatives. The total thorium concentration ranges of interest are:

- less than 10 pCi/gm
- greater than 10 and less than 25 pCi/gm
- greater than 25 and less than 130 pCi/gm
- greater than 130 pCi/gm.

Available Data

Thorium concentrations in soil at this site were measured at 403 borehole locations using a down-hole gamma logging technique. A posting of boring locations is presented in Figure 7.1, with a schematic diagram of the site. At each sampled location on the affected 20-acre portion of the site, a borehole was drilled through the site surface soil, which contains the thorium bearing slag, typically to a depth of about 15 feet. The boreholes were drilled with either 4- or 6-inch diameter augers. Measurements in each borehole were performed starting from the surface and proceeding downward in 6-inch increments.

The primary measurements were made with a 1x1 inch NaI detector (sodium iodide) lowered into the borehole inside a PVC sleeve for protection. One-minute gamma counts were collected (in the integral mode, no energy discrimination) at each position using a "scaler." Gamma counts were converted to thorium 232 (Th_{232}) concentrations in pCi/gm using a calibration algorithm verified with experimental data. The calibration algorithm includes background subtraction and conversion of net gamma counts (counts per minute) to Th_{232} concentration using a semi-empirical detector response function and assumptions regarding the degree of equilibrium between the gamma emitting thorium progeny and Th_{232} in the soil.

The individual gamma logging measurements represent the "average" concentration of Th_{232} (or total thorium as the case may be) in a spherical volume having a radius of approximately 12 to 18 inches. This volume "seen" by the down-hole gamma detector is defined by the effective range in soil of the dominant gamma ray energy (2.6 mev) emitted by thallium 208 (Tl_{208}).

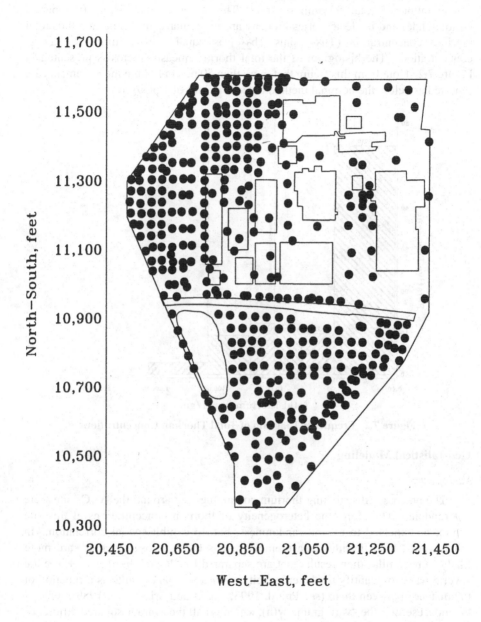

**Figure 7.1   Posting of Bore Hole Locations,
ABC Exotic Metals Site**

The Th_{232} concentration measurements were subsequently converted to total thorium to provide direct comparison to regulatory criteria expressed as concentration of total thorium in soil.   This assumed that Th_{232} (the parent radionuclide) and its decay series progeny are in secular equilibrium and thus total thorium concentration (Th_{232} plus Th_{228}) is equal to two times the Th_{232} concentration.   The histogram of the total thorium measurements is presented in Figure 7.2. Note from this figure that more than 50 percent of the measurements are reported as below the nominal method detection limit of 1 pCi/gm.

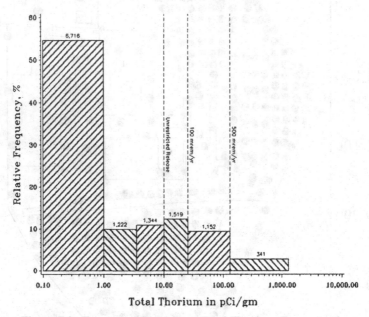

Figure 7.2   Frequency Diagram of Total Thorium Concentrations

Geostatistical Modeling

Variograms

The processes distributing thorium containing slag around the ABCs site were not random.   Therefore, the heterogeneity of thorium concentrations at this site cannot be expected to exhibit randomness, but, to exhibit spatial correlation.   In other words, total thorium measurement results taken "close together" are more likely to be similar than results that are separated by "large" distances.   There are several ways to quantify the heterogeneity of measurement results as a function of the distance between them (see Pitard, 1993; Isaaks and Srivastava, 1989).   One of the most useful is the "variogram," $\gamma(h)$, which is half the average squared difference between paired data values at distance separation h:

$$\gamma(h) \;=\; \frac{1}{2N(h)} \sum_{(i,\,j)\,|\,h_{ii}\,=\,h} (t_i - t_j)^2 \qquad\qquad [7.1]$$

Here N(h) is the number of pairs of results separated by distance h. The measured total thorium data results are symbolized by $t_1, \ldots, t_n$.

Usually the value of the variogram is dependent upon the direction as well as distance defining the separation between data locations. In other words, the difference between measurements taken a fixed distance apart is often dependent upon the directional axis considered. Therefore, given a set of data the values of $\gamma(h)$ maybe be different when calculated in the east-west direction than they are when calculated in the north-south direction. This anisotropic behavior is accounted for by considering "semi-variograms" along different directional axes. Looking at the pattern generated by the semi-variograms often assists with the interpretation of the spatial heterogeneity of the data. Further, if any apparent pattern of spatial heterogeneity can be mathematically described as a function of distance and/or direction, the description will assist in estimation of thorium concentrations at locations where no measurements have been made.

Several models have been proposed to formalize the semi-variogram. Experience has shown the spherical model has proven to be useful in many situations. An ideal spherical semi-variogram is illustrated in Figure 7.3. The formulation of the spherical model is as follows:

$$\Gamma(h) = C_0 + C_1\left[1.5\frac{h}{R} - 0.5\left(\frac{h}{R}\right)^3\right], h < R$$
$$= C_0 + C_1, h \geq R$$
[7.2]

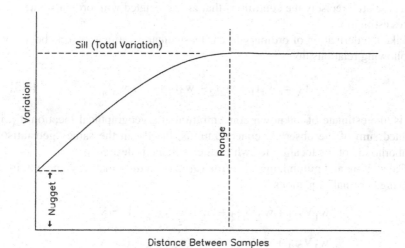

Figure 7.3 Ideal Spherical Model Semi-Variogram

The spherical semi-variogram model indicates that observations very close together will exhibit little variation in their total thorium concentration. This small variation, referred to as the "nugget," C_0, represents sampling and analytical variability, as well as any other source of "random" or unexplained variation. As illustrated in Figure 7.3, the variation between total thorium concentrations can be expected to increase with distance separation until the total variation, $C_0 + C_1$, across the site, or "sill," is reached. The distance at which the variation reaches the sill is referred to as the "range," R. Beyond the range the measured concentrations are no longer spatially correlated.

The practical significance of the range is that data points at a distance greater than the range from a location at which an estimate is desired, provide no useful information regarding the concentration at the desired location. This very important consideration is largely ignored by many popular interpolation algorithms including inverse distance weighting.

Estimation via Ordinary "Kriging"

The important task of estimation of the semi-variogram models is also often overlooked by those who claim to have applied geostatistical analysis by using "kriging" to estimate the extent of soil contamination. The process of "kriging" is really the second step in geostatistical analysis, which seeks to derive an estimate of concentration at locations where no measurement has been made. The desired estimator of the unknown concentration, t_A, should be a linear estimate from the existing data, $t_1, \ldots, t_n$. This estimator should be unbiased in that on the average, or in statistical expectation, it should equal the "true" concentration at that point. And, the estimator should be that member of the class of "linear-unbiased" estimators that has minimum variance (is the "best") about its true value. In other words, the desired kriging estimator is the "best linear unbiased" estimator of the true unknown value, T_A. These are precisely the conditions that are associated with ordinary linear least squares estimation.

Like the derivation of ordinary linear least squares estimators, one begins with the following relationship:

$$t_A = w_1 t_1 + w_2 t_2 + w_3 t_3 + \ldots + w_n t_n \qquad [7.3]$$

That is, the estimate of unknown concentration at a geographical location, t_A, is a weighted sum of the observed concentrations, the t's, in the same "geostatistical neighborhood" of the location for which the estimate is desired.

Calculating and minimizing the error variance in the usual way one obtains the following "normal" equations:

$$w_1 V_{1,1} + w_2 V_{1,2} + \ldots + w_n V_{1,n} + L = V_{1,A}$$

$$w_1 V_{2,1} + w_2 V_{2,2} + \ldots + w_n V_{2,n} + L = V_{2,A}$$

$$\cdot \qquad \cdot \qquad \qquad \cdot \qquad \cdot$$
$$\cdot \qquad \cdot \qquad \qquad \cdot \qquad \cdot \qquad\qquad [7.4]$$
$$\cdot \qquad \cdot \qquad \qquad \cdot \qquad \cdot$$

$$w_1 V_{n,1} + w_2 V_{n,2} + \ldots + w_n V_{n,n} + L = V_{n,A}$$

$$w_1 \qquad + w_2 \qquad + \ldots + w_n \qquad = 1$$

Here $V_{i,j}$ is the covariance between t_i and t_j and, L is the mean of a random function associated with a particular location symbolized by $\vec{x}$. The symbol $\vec{x}$ will be used to designate the three-dimensional location vector (x, y, z).

Geostatistics deal with *random functions*, in addition to random variables. A random function is a set of random variables $\{t(\vec{x}) \mid$ location $\vec{x}$ belongs to the area of interest$\}$ where the dependence among these variables on each other is specified by some probabilistic mechanism. The random function expresses both the random and structured aspects of the phenomenon under study as:

- Locally, the point value $t(\vec{x})$ is considered a random variable.

- The point value $t(\vec{x})$ is also a random function in the sense that for each pair of points $\vec{x}_i$ and $\vec{x}_i + \vec{h}$, the corresponding random variables $t(\vec{x}_i)$ and $t(\vec{x}_i + \vec{h})$ are not independent but related by a correlation expressing the spatial structure of the phenomenon.

In addition, linear geostatistics consider only the first two moments, the mean and variance, of the spatial distribution of results at any point $\vec{x}$. It is therefore assumed that these moments exist and exhibit second-order stationarity. The latter means that (1) the mathematical expectation, $E\{t(\vec{x})\}$, exists and does not depend on location $\vec{x}$; and, (2) for each pair of random variables, $\{t(\vec{x}_i) , t(\vec{x}_i + \vec{h})\}$, the covariance exists and depends only on the separation vector $\vec{h}$.

In this context, the covariances, $V_{i,j}$'s, in the above system of linear equations can be replaced with values of the semi-variograms. This leads to the following system of linear equations for each particular location:

$$
\begin{aligned}
w_1\Gamma_{1,1} + w_2\Gamma_{1,2} + ... + w_n\Gamma_{1,n} + L &= \Gamma_{1,A} \\
w_1\Gamma_{2,1} + w_2\Gamma_{2,2} + ... + w_n\Gamma_{2,n} + L &= \Gamma_{2,A} \\
. \qquad . \qquad\quad . \qquad\quad . \\
. \qquad . \qquad\quad . \qquad\quad . \\
. \qquad . \qquad\quad . \qquad\quad . \\
w_1\Gamma_{n,1} + w_2\Gamma_{n,2} + ... + w_n\Gamma_{n,n} + L &= \Gamma_{n,A} \\
w_1 \quad + w_2 \quad + ... + w_n \quad\quad &= 1
\end{aligned}
$$

$$[7.5]$$

Solving this system of equations for the w's yields the weights to apply to the measured realizations of the random variables, the t's, to provide the desired estimate.

Discussion of the basic concepts and tools of geostatistical analysis can be found in the excellent books by Goovaerts (1997), Isaaks and Srivastava (1989), and Pannatier (1996). These techniques are also discussed in Chapter 10 of the U. S. Environmental Protection Agency (USEPA) publication, *Statistical Methods for Evaluating the Attainment of Cleanup Standards. Volume 1: Soils and Solid Media* (1989).

Journel (1988) describes the advantages and disadvantages of ordinary kriging as follows:

"Traditional interpolation techniques, including triangularization
and inverse distance weighting, do not provide any measure of
the reliability of the estimates .... The main advantage of
geostatistical interpolation techniques, essentially ordinary
kriging, is that an estimation variance is attached to each
estimate .... Unfortunately, unless a Gaussian distribution of
spatial errors is called for, an estimation variance falls short of
providing confidence intervals and the error probability
distribution required for risk assessment.

Regarding the characterization of uncertainty, most interpolation
algorithms, including kriging, are parametric; in the sense that a
model for the distribution of errors is assumed, and parameters of
that model (such as the variance) are provided by the algorithm.
Most often that model is assumed normal or at least symmetric.
Such congenial models are perfectly reasonable to characterize
the distribution of, say, measurement errors in the highly
controlled environment of a laboratory. However they are
questionable when used for spatial interpolation errors ...."

In addition to doubtful distributional assumptions, other problems associated
with the use of ordinary kriging at sites such as the ABC Metals site are:

- How are measurements recorded as below background to be handled in
  statistical calculations? Should they assume a value of one-half background,
  or a value equal to background, or assumed to be zero? (See Chapter 5,
  Censored Data.)

- There are several cases where the total thorium concentrations vary greatly
  with very small changes in depth, as well as evidence that the variation in
  measured concentration is occasionally quite large within small areal
  distances. A series of borings in an obvious area of higher concentration at
  the ABC Metals site exhibit large differences in concentration within an areal
  distance as small as four feet. How these cases are handled in estimating the
  semi-variogram model will have a critical effect on derivation of the
  estimation weights.

Decisions made regarding the handling of measurements less than background
may bias the summary statistics including the sample semi-variograms. The
techniques suggested for statistically dealing with such observations are often
cumbersome to apply (USEPA, 1996) and if such data are abundant may only be
effectively dealt with via nonparametric statistical methods (U.S. Nuclear
Regulatory Commission, 1995). The effect of the latter condition on estimation of
the semi-variogram model is that the "nugget" is apparently equivalent to the sill.
This being the case, the concentration variation at the site would appear to be random
and any spatial structure related to the "occurrence" of high values of concentration

will be masked. If the level of concentration at the site is truly distributed at random, as implied by a semi-variogram with the nugget equal to the sill and a range of zero, then the concentration observed at one location tells us absolutely nothing about the concentration at any other location. An adequate estimate of concentration at any desired location may be simply made in such an instance by choosing a concentration at random from the set of observed concentrations.

Measured total thorium concentrations in the contaminated areas of the site span orders of magnitude. Because the occurrence of high measured total thorium concentration is relatively infrequent, the technique developed by André Journel (1983a, 1983b, 1988) and known as "Probability Kriging" offers a solution to drawbacks of ordinary kriging.

Nonparametric Geostatistical Analysis

Journel (1988) suggests that instead of estimating concentration directly, estimate the probability distribution of concentration measurements at each location.

> "... Non-parametric geostatistical techniques put as a priority, not the derivation of an "optimal" estimator, but modeling of the uncertainty. Indeed, the uncertainty model is independent of the particular estimate retained, and depends only on the information (data) available. The uncertainty model takes the form of a probability distribution of the unknown rather than that of the error, and is given in the non-parametric format of a series of quantiles."

The estimation of the desired probability distribution is facilitated by first considering the empirical cumulative distribution function (ecdf) of total thorium concentration at the site. The ecdf for the observations made at the ABC site is given in Figure 7.4. It is simply constructed by ordering the total thorium concentration observations and plotting the relative frequency of occurrence of concentrations less than the observed measurement. The concept of the ecdf and its virtues was introduced and discussed in Chapter 6.

Note that by using values of the ecdf instead of the thorium concentrations directly, at least two of the major issues associated with ordinary kriging are resolved. The relatively large changes in concentration due to a few high values translate into small changes in the relative frequency that these total thorium concentration observations are not exceeded. If the relative frequency that a concentration level is not exceeded is the subject of geostatistical analysis, instead of the observations themselves, the effect on estimating semi-variogram models of large changes in concentration over small distances is diminished. Thus the resulting estimated semi-variograms are very resistant to outlier data.

Further, issues regarding which value to use for measurements reported as less than background in statistical calculations become moot. All such values are assigned the maximum relative frequency associated with their occurrence. The maximum relative frequency is appropriate because it is the value of a right-continuous ecdf. In

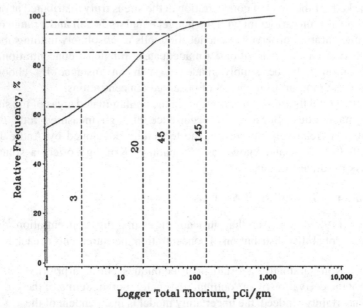

Figure 7.4 Empirical Cumulative Distribution Function Total Thorium

other words, it is desired to describe the cumulative histogram of the data with a continuous curve. To do so it is appropriate to draw such a curve through the upper right-hand corner of each histogram bar.

The desired estimator of the probability distribution of total thorium concentration at any point, $\vec{x}$, is obtained by modeling probabilities for a series of K concentration threshold values T_k discretizing the total range of variation in concentration. This is accomplished by taking advantage of the fact that the conditional probability of a measured concentration, t, being less than threshold T_k is the conditional expectation of an "indicator" random variable, I_k. I_k is defined as having a value of one if t is less than threshold T_k, and a value of zero otherwise.

Four threshold concentrations have been chosen for this site. These are 3, 20, 45, and 145 pCi/gm as illustrated in Figure 7.4. The rationale for choosing precisely these four thresholds is that the ecdf between these thresholds, and between the largest threshold and the maximum measured concentration may be reasonably represented by a series of linear segments. The reason as to why this is desirable will become apparent later in this chapter.

The data are now recoded into four new binary variables, (I_1, I_2, I_3, I_4) corresponding to the four thresholds as indicated above. This is formalized as follows:

$$I_k(\vec{x}) = 1, \text{ if } t(\vec{x}) \le T_k; \ 0, \text{ if } t(\vec{x}) > T_k \qquad [7.6]$$

It is possible to obtain kriged estimators for each of the indicators $I_k(\vec{x})$. The results of such estimation will yield conditional probabilities of not exceeding each of the four threshold concentrations at point $\vec{x}$. These estimates are of the local indicator mean at each location. These estimates are exact in that they reproduce the observed indicator values at the datum locations. However, estimates of the probability of exceeding the indicator threshold are likely to be underestimated in areas of lower concentration and overestimated in areas of higher concentration (Goovaerts, 1997, pp. 293–297). Obtaining "kriged" estimates of the indicators individually ignores indicator data at other thresholds different from that being estimated and therefore does not make full use of the available information.

The additional information provided by the indicators for the "secondary" thresholds can be taken into account by using "cokriging," which explicitly accounts for the spatial cross-correlation between the primary and secondary indicator variables (see Goovaerts, 1997, pp. 185–258). The unfortunate part of indicator cokriging with K indicator variables is that one must infer and jointly estimate K direct and $K(K-1)/2$ cross semi-variograms. If anisotropy is present, meaning that the semi-variogram is directionally dependent, this may have to be done in each of three dimensions. In our present example this translates into 10 direct and cross semi-variograms in each of three dimensions.

Once we have accomplished this feat we then may obtain estimates of the probability that an indicator threshold is, or is not, exceeded that will have theoretically smaller variance than that obtained by using the individual threshold indicators. Goovaerts (1997, pp. 297–300) discusses the virtues and problems associated with indicator cokriging. One of the drawbacks is that when we are finished we only have estimates of the probability that the threshold concentration is, or is not, exceeded at those concentration thresholds chosen. We may refine our estimation by choosing more threshold concentrations and defining more indicators. Thus we may obtain a better definition of the conditional cumulative distribution at the expense of more direct and cross semi-variograms to infer and estimate. This can rapidly become a daunting task.

To make the process manageable, cokriging of the indicator transformed data using the rank-order transform of the ecdf, symbolized by U, as a secondary variable offers a solution. This process is referred to as probability kriging (PK). Goovaerts (1997), Isaaks (1984), Deutsch and Journel (1992), and Journel (1983a, 1983b, 1988) present nice discussions of the nonparametric geostatistical analysis process sometimes referred to as "probability kriging." Other advantages in terms of interpreting the results are discussed by Flatman et al. (1985).

The appropriate PK estimator at point A given the local information in the neighborhood of A is:

$$I_{Ak} = \Sigma_{m=1,n}\lambda_m I_{mk} + \Sigma_{m=1,n}v_m U_m = \text{Prob}[t_A \leq T_k] \quad [7.7]$$

The weights λ_m, v_m are obtained as the solution to the following system of linear equations:

$$\sum_{i=1}^{n} \lambda_{i,k} \Gamma_{I_k,(i,j)} + \sum_{i=1}^{n} v_{i,k} \Gamma_{IU_k,(i,j)} + L_{I_i} = \Gamma_{I_A,(i,A)}, j = 1,\dots, n$$

$$\sum_{i=1}^{n} \lambda_{i,k} \Gamma_{IU_k,(i,j)} + \sum_{i=1}^{n} v_{i,k} \Gamma_{U_k,(i,j)} + L_{U_i} = \Gamma_{IU_A,(i,A)}, j = 1,\dots, n$$

$$\sum_{i=1}^{n} \lambda_{i,k} = 1$$

$$\sum_{i=1}^{n} v_{i,k} = 0 \qquad\qquad [7.8]$$

The above system of equations demands that semi-variograms be established for each of the indicators I_k's, the rank-order transform of the ecdf U, and the covariance between each of the I_k's and U. The sample values of the required semi-variograms are obtained as the following:

$$\gamma_{I_k}(h) = \frac{1}{2N(h)} \sum_{(i,j) \,|\, h_{ij} = h} (I_{k,i} - I_{k,j})^2 \qquad\qquad [7.9]$$

Indicator Semi-Variogram

$$\gamma_U(h) = \frac{1}{2N(h)} \sum_{(i,j) \,|\, h_{ij} = h} (U_i - U_j)^2 \qquad\qquad [7.10]$$

Uniform Transform Semi-Variogram

$$\gamma_{IU_k}(h) = \frac{1}{2N(h)} \sum_{(i,j) \,|\, h_{ij} = h} (I_{k,i} - I_{k,j})(U_i - U_j) \qquad\qquad [7.11]$$

Cross Semi-Variogram

The cross semi-variogram describes the covariance between the indicator variable and the uniform transform variable.

The values of the sample semi-variograms and cross-variograms can be used to estimate the parameters of their corresponding spherical models. These models are as follows for the kth indicator variable:

$$\Gamma_k(h) = C_{I_k,0} + C_{I_k,1}\left[1.5\frac{h}{R_1} - 0.5\left(\frac{h}{R_1}\right)^3\right] + C_{I_k,2}\left[1.5\frac{h}{R_2} - 0.5\left(\frac{h}{R_2}\right)^3\right], h < R_1 < R_2$$

$$= C_{I_k,0} + C_{I_k,1} + C_{I_k,2}\left[1.5\frac{h}{R_2} - 0.5\left(\frac{h}{R_2}\right)^3\right], R_1 < h < R_2 \qquad\qquad [7.12]$$

$$= C_{I_k,0} + C_{I_k,1} + C_{I_k,2}, R_1 < R_2 < h$$

The model for the uniform transformation variable is:

$$\Gamma_U(h) = C_{U,0} + C_{U,1}\left[1.5\frac{h}{R_1} - 0.5\left(\frac{h}{R_1}\right)^3\right] + C_{U,2}\left[1.5\frac{h}{R_2} - 0.5\left(\frac{h}{R_2}\right)^3\right], h < R_1 < R_2$$

$$= C_{U,0} + C_{U,1} + C_{U,2}\left[1.5\frac{h}{R_2} - 0.5\left(\frac{h}{R_2}\right)^3\right], R_1 < h < R_2 \qquad [7.13]$$

$$= C_{U,0} + C_{U,1} + C_{U,2}, R_1 < R_2 < h$$

For the cross-variograms the models are defined as:

$$\Gamma_{UI_k}(h) = C_{UI_k,0} + C_{UI_k,1}\left[1.5\frac{h}{R_1} - 0.5\left(\frac{h}{R_1}\right)^3\right] + C_{UI_k,2}\left[1.5\frac{h}{R_2} - 0.5\left(\frac{h}{R_2}\right)^3\right],$$

$$h < R_1 < R_2$$

$$= C_{UI_k,0} + C_{UI_k,1} + C_{UI_k,2}\left[1.5\frac{h}{R_2} - 0.5\left(\frac{h}{R_2}\right)^3\right], R_1 < h < R_2 \quad [7.14]$$

$$= C_{UI_k,0} + C_{UI_k,1} + C_{UI_k,2}, R_1 < R_2 < h$$

Note that these models contain two ranges, R_1 and R_2, and associated sill coefficients, C_1 and C_2, reflecting the presence of two plateaus suggested by the sample semi-variograms. This representation defines a "nested" structural model for the semi-variogram. The sample and estimated models for semi-variograms are presented in the Figures 7.5–7.8. The estimated semi-variogram model is represented by the continuous curve, and the sample semi-variogram is represented by the points shown in these figures.

There are 27 semi-variograms appearing in Figures 7.5–7.8. Because of the geometric anisotropy indicated by the data, nine variograms are required in each of three directions. These nine semi-variograms are distributed as one for the uniform transformed data, four for the indicator variables and four cross semi-variograms between the uniform transform and each of the indicator variables.

The derivation of the semi-variogram models employed the software of GSLIB (Deutsch and Journel, 1992) to calculate the sample semi-variograms and SAS/Stat (SAS, 1989) software to estimate the ranges and structural coefficients of the semi-variogram models. Estimation of the structural coefficients, i.e., the nugget and sills, involves nonlinear estimation procedures constrained by the requirements of coregionalization. This simply means that the semi-variogram structures for an indicator variable, that for the uniform transform and their cross semi-variogram must be consonant with each other. Coregionalization demands that coefficients $C_{I,m}$ and $C_{U,m}$ be greater than zero, for all $m = 0, 1, 2$, and that the following determinant be positive definite:

$$\begin{vmatrix} C_{I,m} & C_{UI,m} \\ C_{UI,m} & C_{U,m} \end{vmatrix} \qquad [7.15]$$

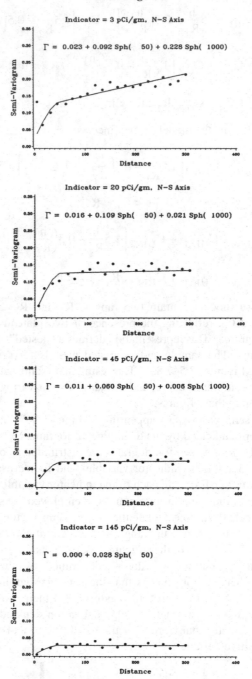

Figure 7.5A N-S Indicator Semi-variograms

Cross Semi-variogram

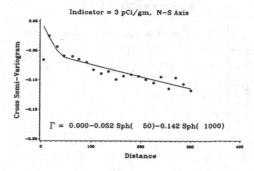

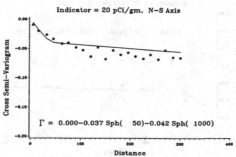

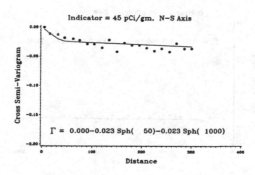

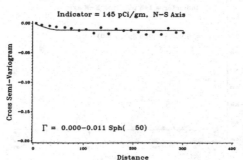

Figure 7.5B   N-S Indicator Semi-variograms

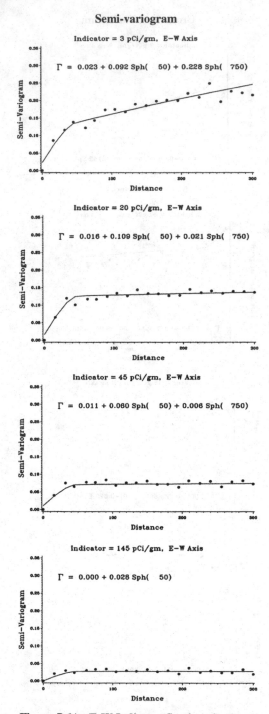

Figure 7.6A   E-W Indicator Semi-variograms

Cross Semi-variogram

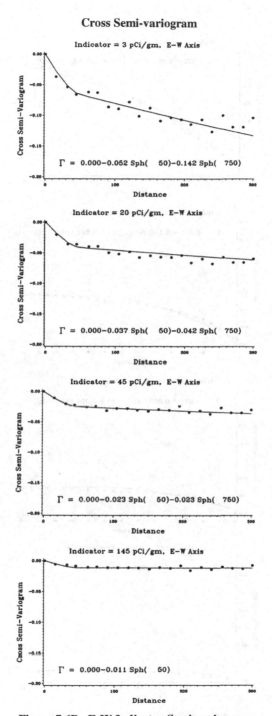

Figure 7.6B E-W Indicator Semi-variograms

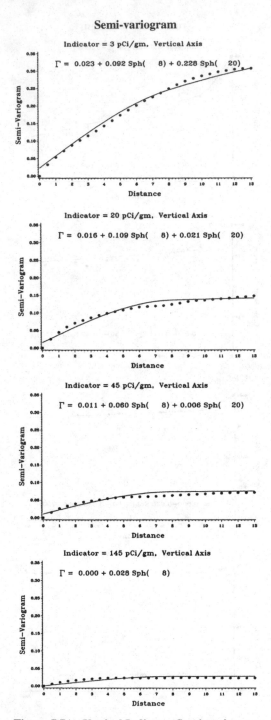

Figure 7.7A Vertical Indicator Semi-variograms

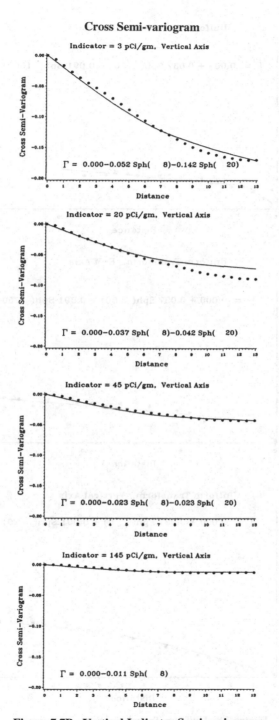

Figure 7.7B Vertical Indicator Semi-variograms

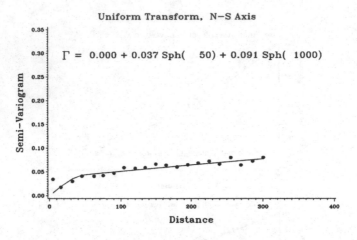

Uniform Transform, N–S Axis

$\Gamma = 0.000 + 0.037\ \text{Sph}(\quad 50) + 0.091\ \text{Sph}(\ 1000)$

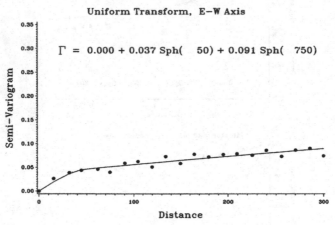

Uniform Transform, E–W Axis

$\Gamma = 0.000 + 0.037\ \text{Sph}(\quad 50) + 0.091\ \text{Sph}(\quad 750)$

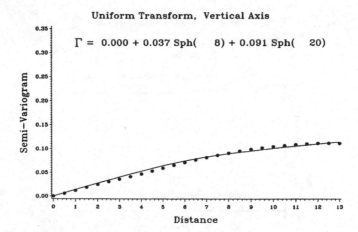

Uniform Transform, Vertical Axis

$\Gamma = 0.000 + 0.037\ \text{Sph}(\qquad 8) + 0.091\ \text{Sph}(\qquad 20)$

Figure 7.8   Uniform Transform Semi-variograms

The coregionalization requirements lead to the following practical rules:

- Any structure that appears in the cross semi-variogram must also appear in both the indicator and uniform semi-variograms.

- A structure appearing in either the indicator or uniform semi-variograms does not necessarily have to appear in the cross semi-variogram model.

While one might argue that some of the semi-variogram models do not appear to fit the sample semi-variograms very well, in practice an assessment must be made as to whether improving the fit of all semi-variograms is worth the effort. If by doing so has little effect on estimation and significantly complicates specification of the kriging model it probably is not worth the additional effort. Such was the judgment made here.

Some Implications of Variography

The semi-variograms provide some interesting information regarding the spatial distribution of total thorium at the site. The semi-variogram models for the first three indicators and the uniform transform are described by variogram models (see Figures 7.5–7.8) with an isotropic nugget effect and two additional transition structures as described above. The first of the transition structures exhibits a range of 50 feet and is isotropic in the horizontal (x, y) plane but shows directional anisotropy between the vertical (z) and the x, y plane. The second transition structure exhibits directional anisotropy among all directions. Directional anisotropy is characterized by constant sill coefficients, C_1 and C_2, and directionally dependent ranges for each transition structure.

The directional anisotropy in the x, y plane is interesting. The estimated range along the north-south axis is 1,000 feet, but only 750 feet in the east-west direction. This suggests that where low to moderate thorium concentrations are found, they will tend to occur in elliptical regions with the long axis oriented in a north-south direction. This orientation is consonant with the facility layout and traffic patterns.

The semi-variogram models for the 145 pCi/gm indicator appear to be isotropic in the x, y plane and exhibit only one transition structure. The range of this transition structure is estimated to be 50 feet. This suggests that when high concentrations of total thorium are found, they tend to define rather confined areas. Note that this indicator exhibits one transition structure in the vertical direction. This suggests that the horizontally confined areas of high concentration tend to be confined in the vertical direction as well.

There are tools other than the semi-variogram for investigating the relationship among observations as a function of their distance separation. These include the "Standardized Variogram," the "Correlogram," and the "Madogram." These are mentioned here without definition to recognize their existence. Their definition is not necessary to our discussion as to why one needs to pay attention to the structure of the spatial relationships among observations. The interested reader is referred to Goovaerts (1997), Isaaks and Srivastava (1989), and Pannatier (1996), among others, for a complete discussion of these tools.

In addition to providing insight into the deposition process and possible changes in the deposition process that might occur for different concentration ranges, variography also permits an assessment of sampling adequacy. It was mentioned earlier that observations at a distance away from a point of interest that is greater than the range provide no information about the point of interest. Thus the ranges associated with the directional semi-variograms define an ellipsoidal "neighborhood" about a point of interest in which we may obtain information about the point of interest. A practical rule-of-thumb is that this neighborhood is defined by axes equivalent to two-thirds the respective range.

Looking at the collection of available samples, should we find that there are no samples within the neighborhood of a point of interest, then the existing collection of samples is inadequate. We have also indicated the physical locations for the collection of additional samples. The interpolation algorithms often found with Geographical Information System (GIS), including the popular inverse-weighted distance algorithm, totally ignore the potential inadequate sampling problem.

Estimated Distribution of Total Thorium Concentration

Once the semi-variogram models were obtained they were used to estimate the conditional probability distribution of total thorium concentration at the centroid of 253,344 blocks across the site. Kriged estimates were obtained for each block of dimension 2.5 m by 2.5 m by 0.333m (8.202 ft by 8.202 ft by 1.094 ft) to a nominal depth of 15 feet. The depth restriction is imposed because only a very few borings extend to, or beyond, that depth. All measured concentrations beyond a depth of 15.5 feet are recorded as below background. Each block is oriented according to the usual coordinate axes.

Truly three-dimensional PK estimation was performed to obtain the conditional probability that the total thorium concentration will not exceed each of the four indicator concentrations. This estimation employed PK software developed specifically for Splitstone & Associates by Clayton Deutsch, Ph.D., P.E. (1998) while at Stanford University. PK estimation was restricted to use up to 8 nearest data values within an elliptical search volume centered on the point of estimation. The principal axes of this elliptical region were chosen as 670 ft, 500 ft, and 10 ft in the principal directions. The lengths of these axes correspond to approximately two-thirds of the effective directional ranges. During semi-variogram estimation it was concluded that no rotation of the principal axes from their usual directions was necessary.

Upon completion of PK, the grid network of points of estimation was restricted to account for the irregular site boundary and other salient features such as buildings and roads existing prior to the production of the ferrocolumbium alloy. It makes logical sense to impose this restriction on the mathematical estimation as the thorium slag is not mobile. Figure 7.9 shows the areal extent of block grid after applying restrictions. This grid defines the areal centroid locations of 152,124 "basic remedial blocks" of volume 2.7258 cubic yards (cu yds).

The results immediately available upon completion of PK estimation are the conditional probabilities that the total thorium concentration will not exceed each of the four indicator concentrations at each point of estimation. Because the basic

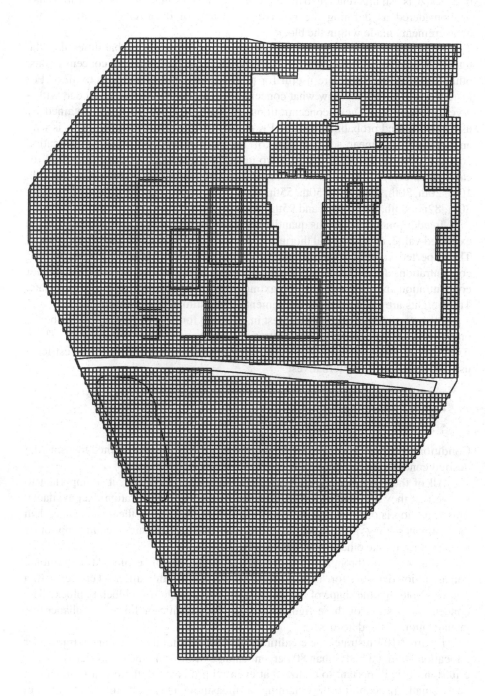

Figure 7.9   PK Estimation Grid Schematic

blocks size is "small" relative to the majority of data spacing, these probabilities may be considered as defining the relative frequency of occurrence of all possible measurements made within the block.

While it certainly is beneficial to know the conditional probabilities that the total thorium concentration will not exceed each of the four indicator concentrations, other statistics may be more useful for planning decommissioning activities. For instance, it is useful to know what concentration levels will not be exceeded with a given probability. These concentrations, or quantiles, can be easily obtained by using the desired probability and the PK results to interpolate the ecdf. This is why an approximate linear segmentation of the ecdf when choosing the indication concentrations is of value. Twenty-two quantile concentrations were estimated for each block corresponding to the following percentiles of the distribution: the 5th, 10th, 20th, 30th, 40th, 45th, 50th, 55th, 60th, 63th, 65th, 67th, 70th, 73th, 75th, 77th, 80th, 82nd, 85th, 87th, 90th, and 95th.

In addition to the various quantile estimates of total thorium concentration, the expected value, or mean, total thorium concentration for any block may be obtained. The expected value is easily calculated as the weighted average of the mean ecdf concentrations found between the indicator concentration values, or an indicator concentration and the minimum or maximum observed concentration as appropriate. The weights are supplied by the incremental PK estimated probabilities.

Other statistics of potential interest in planning for decommissioning may be the probabilities that certain fixed concentrations are exceeded (or not exceeded). These fixed concentrations are defined by the NRC dose criteria for release for unrestricted use and restricted-use alternatives:

- greater than 10 pCi/gm
- greater than 25 pCi/gm, and
- greater than 130 pCi/gm.

Conditional estimates of the desired probabilities can be easily obtained by using the desired concentrations and the PK results to interpolate the ecdf.

All of these estimates are labeled "conditional." However, it is important to realize that they are conditional only on the measured concentration data available. This condition is one that applies to any estimation method. If the data change, then the estimates may also change. Nonparametric geostatistics require no other assumptions, unlike other estimation techniques.

Figures 7.10 through 7.14 present a depiction of estimated conditional concentration densities for "typical" basic blocks that fall into different concentration ranges. Note that the shape of the distribution will change from block to block. The concentration scale of these figures is either logarithmic or linear to enhance the visualization of the densities.

Figure 7.10 illustrates the conditional density of total thorium concentration for a location having a better than 80 percent chance of meeting the unrestricted release criterion. It is important to realize that even with this block there is a finite, albeit very small, probability of obtaining a measurement result for total thorium exceeding 130 pCi/gm.

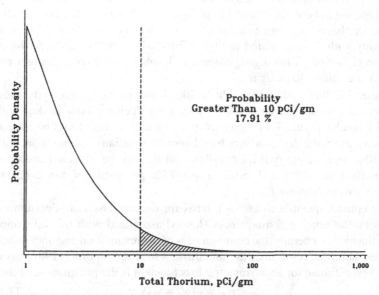

Figure 7.10 "Typical" Density Block Type 1

Figure 7.11 illustrates the conditional density for a "typical" block, which may be classified in concentration range between 10 and 25 pCi/gm. Note that for this block there is a better than 60 percent chance of a measured concentration being below 25 pCi/gm.

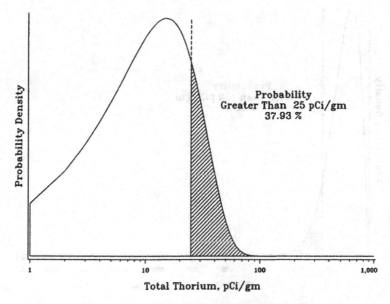

Figure 7.11 "Typical Density Block Type 2

Figure 7.12 illustrates the shape of a "typical" block concentration density for a block classified as between 25 and 130 pCi/gm. For this block there is a greater than 90% chance that a measured concentration will be less than 130 pCi/gm. Contrast this density with that illustrated in Figure 7.13, which is "typical" of a block that would be classified as having significant soil volume with total thorium concentration greater than 130 pCi/gm.

These densities will have a high likelihood of being correctly classified regardless of the statistic one chooses to use for deciding how to deal with that particular block. Figure 7.14, however, presents a case that would be released for unrestricted use if the decision were based upon the median concentration, put in the 25 to 130 pCi/gm category if the decision statistic was the expected concentration, and classified for off-site disposal if the 75th percentile of the concentration distribution were employed.

The optimal quantile to use as a basis for decommissioning decision making depends on the relative consequences (losses) associated with overestimation and underestimation. Generally, the consequences of overestimation and underestimation are not the same and result in an asymmetric loss function. Journel (1988) has shown that the best estimate for an asymmetric loss function is the pth quantile of the cdf:

$$p = w_2/(w_1 + w_2)$$
[7.16]

where:

w_1 = cost of overestimation per yd^3
w_2 = cost of underestimation per yd^3.

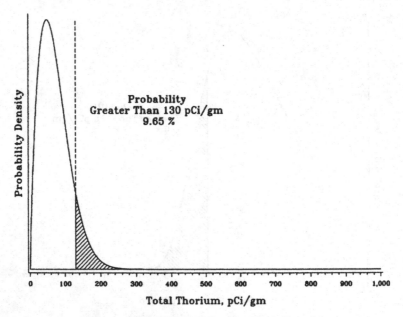

Figure 7.12 "Typical" Density Block Type 3

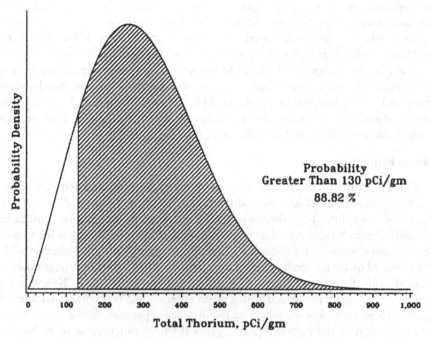

Probability
Greater Than 130 pCi/gm
88.82 %

Figure 7.13 "Typical" Density Block Type 4

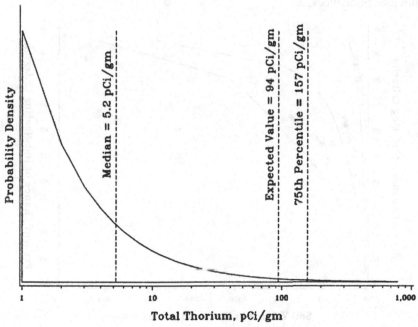

Figure 7.14 "Typical" Density Block Type 5

The specification of the unit costs, w_1 and w_2, reflects the perceived losses associated with wrong decisions. Srivastava (1987) provides a very nice discussion of asymmetric loss functions. Usually, w_1 is easily specified. It is the same as the cost of remediating any unit, only now it applies to essentially clean soil.

The cost w_2 is more difficult to determine. It is, at first glance, the cost of leaving contaminated material *in situ*. Some would argue that this cost then becomes infinite as they are tempted to add the cost of a life, etc. However, because of the usual confirmatory sampling, w_2 is really the cost of disposal plus perhaps remobilization and additional confirmatory sampling.

Volume Estimation

Often it is difficult to elicit the costs associated with over- and underestimation and, therefore, an optimal decision statistic for block classification. A distribution of possible of soil volume to be dealt with during decommissioning can be constructed by classifying each block to a decommissioning alternative based upon the various total thorium concentration quantiles. Classifying each block to an alternative by each of the 22 quantiles sequentially, a distribution of soil volume estimates may be obtained. This distribution is presented in Table 7.1 and Figure 7.15. Note that it is the volume increment associated with a given criterion that is reported in Table 7.1. Figure 7.15 presents this increment as the distance separation between successive curves. Therefore, the rightmost curve gives the total soil volume to be handled during decommissioning. Illustrated by a dashed line is the case in which the volume determination decision rule is applied to the expected value of concentration within each basic block.

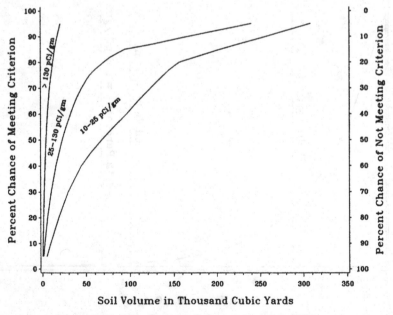

Figure 7.15 **Distribution of Contaminated Soil Volume**

Table 7.1
Contaminated Soil Volume Estimates,
Decision Based upon Estimated Concentration Percentile
(Individual 2.5 Meter x 2.5 Meter x 0.333 Meter Blocks)

Decision Percentile	Soil/Slag Volume Estimates in Cubic Yards			
	Greater Than 130 pCi/gm	25–130 pCi/gm	10–25 pCi/gm	Clean (Less Than 10 pCi/gm)
5	368	842	3,707	409,751
10	594	1,856	6,651	405,567
20	1,139	4,427	12,719	396,384
30	1,922	7,722	19,820	385,205
40	2,775	12,427	29,431	370,036
45	3,336	15,407	36,467	359,459
50	3,972	18,495	45,181	347,021
55	4,738	22,265	54,307	333,359
60	5,640	26,460	63,943	318,626
63	6,239	29,172	68,444	310,814
65	6,692	31,157	71,259	305,561
67	7,210	33,187	74,315	299,956
70	8,044	36,794	78,268	291,564
73	9,012	41,090	81,833	282,734
75	9,846	44,153	84,106	276,563
77	10,571	49,071	84,856	270,171
80	11,718	58,047	86,459	258,444
82	12,629	64,717	97,499	239,824
85	14,019	80,009	110,133	210,507
87	15,011	109,683	99,745	190,230
90	16,740	149,080	91,325	157,525
95	20,267	219,598	68,326	106,478

More About Variography

Frequently, the major objective of data summarization for a site becomes a rush to translate data into neatly colored areas enclosed by concentration isopleths. Unfortunately, this objective is aided and abetted by various popular GIS packages. Some of these packages allege that estimation via "kriging" is an option. While this may be technically true, none of these packages emphasize the importance of first investigating the semi-variogram models that drive the estimation via "kriging." In the author's opinion, investigating the spatial structure of the data via variography is the most important task of those seeking to make sense of soil sampling data. In many cases, the patterns of spatial relationships revealed by sample semi-variograms provide interesting insights without the need for further estimation. Following are two examples.

Extensive soil sampling and assay for metals has occurred in the vicinity of the ASARCO Incorporated Globe Plant in Denver, Colorado. Almost 22,500 soil samples have been assayed for arsenic within a rectangular area 3.39 kilometers in north-south dimension and 2.12 kilometers in east-west dimension. The ASARCO Globe Plant located in the northwestern quadrant of this area has been the site of various metal refining operations since 1886. Gold and silver were refined at this facility until 1901 when the facility was converted to lead smelting. In 1919 the plant was converted to produce refined arsenic trioxide. Arsenic trioxide production terminated in 1926.

Semi-variograms for indicator variables corresponding to selected arsenic concentrations and the rank-order (uniform) transformed data were constructed along the north-south and east-west axes. These are presented in Figure 7.16. Note that the sill is reached within a few hundred meters for all the semi-variograms. Some interesting structural differences occur between the north-south and east-west semi-variograms as the indicator concentrations increase. Keeping in mind that smaller values of the semi-variogram indicate similarity, note that the semi-variogram in the east-west direction start to decease between 500 and 1,000 meters for indicator variables representing concentrations in excess of 10 mg/kg. This is not true for the semi-variograms along the north-south axis. This suggests that deposition of arsenic lies within a band along the north-south axis and a width of perhaps 1,500 meters in the east-west direction. Because this width is somewhat less than the east-west dimension of the sampled area, it is unlikely that this behavior is an artifact of the sampling scheme.

The indicator semi-variogram is a valuable tool in conducting environmental investigations and for providing insight to patterns of increased concentration. Just to the east of the Globe Plant site is a site known as the "Vasquez Boulevard and I-70 Site." One of the reasons this site is interesting is that Region 8 of the USEPA in cooperation with the City and County of Denver and the Colorado Department of Public Health and Environment have conducted extensive sampling of surface soil on a number of residential properties.

One of the stated objectives of this study was to investigate the potential for spatial patterns of metals concentrations in the soil at a single residential site. Figure 7.17 presents the schematic map of sampling locations at Site #3 in this study. Two hundred and twenty-four (224) samples were collected on a grid with nominal 5-foot spacing.

N-S Axis   E-W Axis

a. 10-mg/kg Indicator

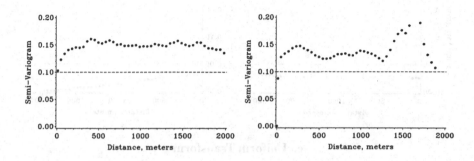

b. 35-mg/kg Indicator

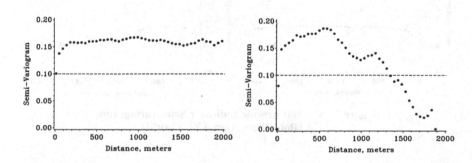

c. 72-mg/kg Indicator

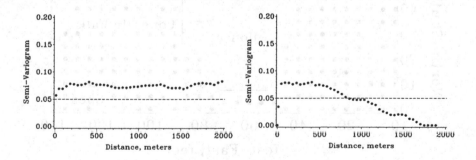

Figure 7-16   Soil Arsenic Indicator Semi-variograms,
Globe Plant Area, CO

N-S Axis   **E-W Axis**

d. 179-mg/kg Indicator

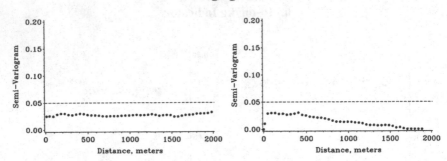

e. Uniform Transform

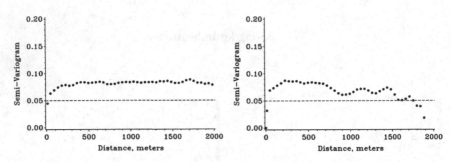

Figure 7-16   Soil Arsenic Indicator Semi-variograms,
                Globe Plant Area, CO (Cont'd)

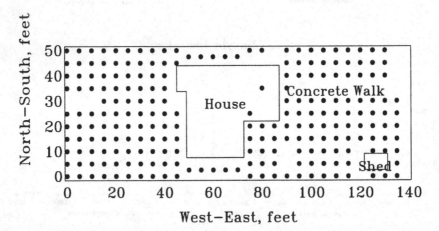

Figure 7-17   Sampling Locations,
                Residential Risk-Based Sampling Site #3 Schematic,
                Vasquez Boulevard and I-70 Site, Denver, CO

Semi-variograms for selected indicator variables corresponding to arsenic levels of 70 and 300 mg/kg are shown in Figure 7.18. Note that these sample semi-variograms clearly indicate the existence of a spatial pattern among elevated arsenic concentrations. The implication that spatial correlation of contaminant concentration does exist in areas the size of a typical residential lot has some interesting implications. Not the least of these is that random exposure exists only in the mind of the risk assessor who is left to assume that an individual moves randomly around the property. The authors (Ginevan and Splitstone, 1997) have suggested an alternative approach using probability of encountering a concentration level at each site location coupled with the probability that a particular location will be visited. This permits the evaluation of various realistic exposure scenarios.

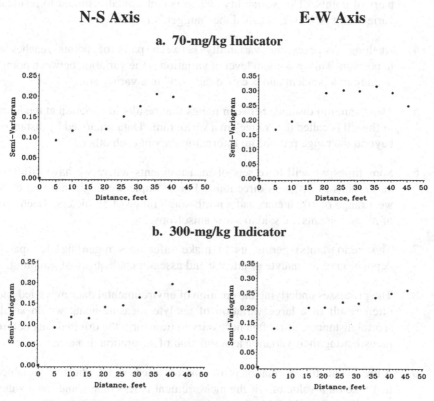

N-S Axis   **E-W Axis**

a. 70-mg/kg Indicator

b. 300-mg/kg Indicator

Figure 7-18   Soil Arsenic Indicator Semi-variograms,
Residential Risk-Based Sampling Site #3 Schematic,
Vasquez Boulevard and I-70 Site, Denver, CO

A Summary of Geostatistical Concepts and Terms

The material discussed in this chapter is well beyond the statistics usually encountered by the average environmental manager. Hopefully, those who find themselves in that position will have taken the advice given in the preface and have read this chapter ignoring the mathematical formulae. The following is a brief summary of the concepts discussed.

1. Environmental measurements taken some distance from each other are less similar than those taken close together.

2. The relative disparity among measurements as a function of their separation distance may be described by one-half the average sum of squares of their differences calculated at various separation distances. The results of these calculations are summarized in the variogram.

3. At very small separation distances there is still some variability between pairs of points. This variability, which is conceptually similar to residual variation in ANOVA, is called the "nugget" in a variogram.

4. At long distances the variability between pairs of points reaches a maximum. This maximum level of variation is the variation between points that are independent and is called the "sill" in a variogram.

5. The minimum distance between points that results in variation at the level of the sill is called the "range" in a variogram. Data separated by distances beyond the range provide no information about each other.

6. Sometimes we will have sets of measurements where we have different variograms in different directions. For example we might have an east-west range of 100 meters and a north-south range of 50 meters. Such sets of measurements are said to show anisotropy.

7. The variogram(s) permit us to make inferences regarding the spatial deposition of an analyte of interest and assess the adequacy of sampling.

8. The processes underlying the creation of environmental data are varied and often result in a large variation of analyte measurements within small spatial distances. It is often necessary to transform the original data before investigating their variation as a function of separation distance.

9. A useful data transformation is to construct a series of "indicator" variables that take on a value of 1 if the measurement is less than some fixed value and 0 if it is less than this fixed value.

10. Variograms of the transform data can be used to assess the adequacy of the existing sampling effort to make the decisions required.

11. We can also use the variogram(s) of the transformed data to develop equations to predict probability that a fixed concentration is exceeded by unknown points. This prediction process is called "kriging" after Daniel Krige, a South African mining engineer who developed the technique.

12. Kriging, applied to these indicator variables is called indicator kriging. The objective is to predict the probability that a given point will be below or above some threshold.

13. If we form a number of indicator variables for a number of thresholds and then do cokriging with the rank-order transformation of the empirical cumulative distribution function, the result is called probability kriging, and can provide an estimate of the cumulative distribution of concentrations at a given point.

14. We prefer the nonparametric kriging methods (indicator and probability kriging) because they require no distributional assumptions and because they are less influenced by very large or small sample values, and issues with regard to values reported as below the method limit of detection are moot.

Epilogue

The examples have been chosen to illustrate the concepts and techniques of nonparametric geostatistics. The reader is cautioned that in the experience of the authors they are atypical in terms of the volume of available data. It has been the authors' experience that most industrial sites are close to 10 acres in size and have 100, or fewer, sampling locations spread over a variety of depths. Further, these sampling locations are not generally associated with any regular spatial grid, but are usually selected by "engineering judgment." While these conditions make the geostatistical analysis more difficult in terms of semi-variogram estimation, it can be successfully accomplished. These cases just require a little more effort and increased "artistic" sensibility conditioned with experience.

We would be remiss if we didn't recognize that simpler tools often provide useful answers. Two such tools are posting plots and Thiessen polygons.

A *posting plot* is simply a two-dimensional plot of the sampling locations. Figures 7.1 and 7.17 are posting plots. The utility of the posting plot is increased by labeling the plotted points to show the value of analyte concentration. Sometimes one sees posting plots where the actual measured value(s) are printed for each point. However, such plots are difficult to read and it is our view that coding the points by either color or symbol produces a more useful graphic. Figure 7.19 shows a hypothetical posting plot. Here we have plotted the four quartiles of the data with symbols: ● ■ ▲ ✦ . In this plot we can easily see that the sample points form a regular grid and that there is a north-south gradient in the concentration, with lower concentrations to the north and higher concentrations to the south.

Thiessen polygons result from a "tessellation" of the plane sampled, called a bounded Voronoi diagram (Okabe et al., 2001). Tessellation is covering of a plane surface by congruent geometric shapes so that there are no gaps or overlaps. Details of constructing a bounded Voronoi diagram are provided in Okabe et al. (2001). A very simple Voronoi diagram is shown in Figure 7.20. The important properties of a Voronoi diagram are that if one begins with N sample points, the Voronoi diagram divides the parcel of interest into exactly N polygons (which we term Thiessen

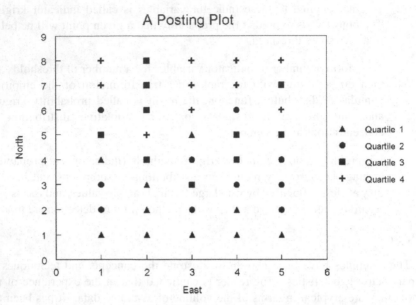

Figure 7.19   A Posting Plot of Some Hypothetical Data
(Note that there appears to be a north-south gradient.)

polygons after Thiessen, 1911, with each polygon containing one and only one sample point. For each sample point, any point inside the polygon is closer to the sample point in the polygon than it is to any other sample point. Thus, in a sense, the area of the polygon associated with a given sample point, which we can think of as the "area of influence" of the sample point, is a local measure of sample density. That is, sampling density is usually thought of as points per unit area, but the Thiessen polygons give us the area related to each sample point, which is the reciprocal of the usual density measure.

These properties can be useful. First, we can use the areas of the Thiessen polygons as weights to calculate weighted mean concentrations for environmental exposure assessments (Burmaster and Thompson, 1997). The rationale for such a calculation is that if all points inside the polygon are closer to the sample point inside the polygon than they are to any other sample point, it makes sense to represent the concentration in the polygon with the concentration measured for the sample point in the polygon. Moreover, if a person is randomly using the environment, the fraction of his/her exposure attributable to the ith polygon should be equal to the area of the ith polygon divided by the total area of the parcel being assessed, which is also the sum of the area of all Thiessen polygons.

Another way in which Thiessen polygons can be useful is in determining whether sampling has tended to concentrate on dirty areas. That is, one can plot sample concentration against the area of the Thiessen polygon associated with that sample. If it looks like large sample values are associated with small polygon areas, it provides evidence that high sample values are associated with areas of high sample density (and thus small polygons) and thus that highly contaminated areas tend to be sampled more often than less contaminated areas. One can test such apparent associations by calculating Spearman's coefficient of rank correlation (Chapter 4) for sample value and polygon area. A significant negative correlation would provide evidence that high concentration areas were oversampled.

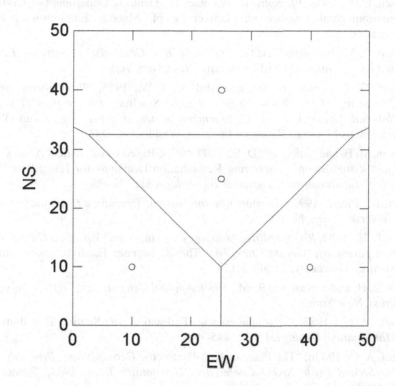

Figure 7.20 A Voronoi Tesselation on Four Data Points
(The result is 4 Thiessen polygons.)

References

Burmaster, D. S. and Thompson, K. M., 1997, "Estimating Exposure Point Concentrations for Surface Soil for Use in Deterministic and Probabilistic Risk assessments," *Human and Ecological Risk Assessment*, 3(3): 363–384.

Deutsch, C. V., 1998, *PK Software* (Contact Dr. Deutsch, Department of Civil and Environmental Engineering, University of Alberta, Edmonton, Alberta, Canada).

Deutsch, C. V. and Journel, A. G., 1992, *GSLIB — Geostatistical Software Library and User's Guide*, Oxford University Press, New York.

Flatman, G. T., Brown, K. W., and Mullins, J. W., 1985, "Probabilistic Spatial Contouring of the Plume Around a Lead Smelter," *Proceedings of the 6th National Conference on Management of Uncontrolled Hazardous Waste*, Hazardous Materials Research Institute, Washington, D.C.

Ginevan, M. E. and Splitstone, D. E., 1997, "Risk-Based Geostatistical Analysis and Data Visualization: Improving Remediation Decisions for Hazardous Waste Sites," *Environmental Science & Technology*, 31: 92–96.

Goovaerts, Pierre, 1997, *Geostatistics for Natural Resources Evaluation*, Oxford University Press, New York.

Isaaks, E. H., 1984, *Risk Qualified Mappings for Hazardous Wastes, a Case Study in Non-parametric Geostatistics*, MS Thesis, Branner Earth Sciences Library, Stanford University, Stanford, CA.

Isaaks, E. H. and Srivastava, R. M., 1989, *Applied Geostatistics*, Oxford University Press, New York.

Journel, A. G., 1983a, "Nonparametric Estimation of Spatial Distributions," *Mathematical Geology*, 15(3): 445–468.

Journel, A. G., 1983b, "The Place of Non-Parametric Geostatistics," *Proceedings of the Second World NATO Conference 'Geostatitics Tahoe 1983,'* Reidel Co., Netherlands.

Journel, A. G., 1988, "Non-parametric Geostatistics for Risk and Additional Sampling Assessment," *Principles of Environmental Sampling*, ed. L. Keith, American Chemical Society, pp. 45-72.

Okabe, A., Boots, B., Sugihara, K., and Chiu, S. N., 2001, *Spatial Tesselations: Concepts and Applications of Voronoi Diagrams (Second Edition)*. John Wiley, New York.

Pannatier, Y., 1996, *VARIOWIN Software for Spatial Data Analysis in 2D*, Springer-Verlag, New York, ISBN 0-387-94679-9.

Pitard, F. F., 1993, *Pierre Gy's Sampling Theory and Sampling Practice, Second Edition*, CRC Press, Boca Raton, FL.

SAS, 1989, *SAS/STAT User's Guide, Version 6, Fourth Edition, Volume 2*, SAS Institute Inc., Cary, NC.

Srivastava, R. M., 1987, "Minimum Variance or Maximum Profitability," *Canadian Institute of Mining Journal*, 80(901).

Thiessen, A. H., 1911, "Precipitation Averages for Large Areas," *Monthly Weather Review*, 39: 1082–1084.

USEPA, 1989, *Methods for Evaluating the Attainment of Cleanup Standards. Volume 1: Soils and Solid Media*, Washington D.C., EPA 230/02-89-042.

USEPA, 1996, *Guidance for the Data Quality Assessment, Practical Methods for Data Analysis, EPA QA/G-9, QA96 Version*, EPA/600/R-96/084.

U. S. Nuclear Regulatory Commission, 1995, *A Nonparametric Statistical Methodology for the Design and Analysis of Final Status Decommissioning Surveys*. NUREG-1505.

CHAPTER 8

Tools for the Analysis of Temporal Data

"In applying statistical theory, the main consideration is not what the shape of the universe is, but whether there is any universe at all. No universe can be assumed, nor ... statistical theory ... applied unless the observations show statistical control."

"... Very often the experimenter, instead of rushing in to apply [statistical methods] should be more concerned about attaining statistical control and asking himself whether any predictions at all (the only purpose of his experiment), by statistical theory or otherwise, can be made." (Deming, 1950)

All too often in the rush to summarize available data to derive indices of environmental quality or estimates of exposure, the assumption is made that observations arise as a result of some random process. Actually, experience has shown that the statistical independence of environmental measurements at a point of observation is a rarity. Therefore, the application of statistical theory to these observations, and resulting inferences are simply not correct.

Consider the following representation of hourly concentrations of airborne particulate matter less than 10 microns in size (PM_{10}) made at the Liberty Borough monitoring site in Allegheny County, Pennsylvania, from January 1 through August 31, 1993.

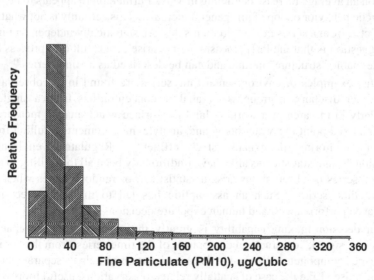

Figure 8.1   Hourly PM_{10} Observations,
Liberty Borough Monitor, January–August 1993

The shape of this frequency diagram of PM_{10} concentration is typical in air quality studies, and popular wisdom frequently suggests that these data might be described by the statistically tractable log-normal distribution. However, take a look at this same data plotted versus time.

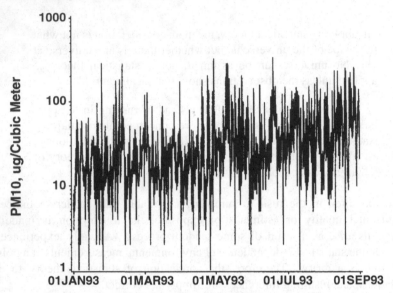

Figure 8.2   Hourly PM_{10} Observations versus Time,
Liberty Borough Monitor, January–August, 1993

Careful observation of this figure suggests that the concentrations of PM_{10} tend to exhibit an average increase beginning in May. Further, there appears to be a short-term cyclic behavior on top of this general increase. This certainly is not what would be expected from a series of measurements that are statistically independent in time. The suggestion is that the PM_{10} measurements arise as a result of a process having some definable "structure" in time and can be described as a "time series."

Other examples of environmental time series are found in the observation of waste water discharges; groundwater analyte concentrations from a single well, particularly in the area of a working landfill; surface water analyte measurements made at a fixed point in a water body; and, analyte measurements resulting from the frequent monitoring of exhaust stack effluent. Regulators, environmental professionals, and statisticians alike have traditionally been all too willing to assume that such series of observations arise as statistical, or random, series when in fact they are time series. Such an assumption has led to many incorrect process compliance performances, and human exposure decisions.

Our decision-making capability is greatly improved if we can separate the underlying "signal," or structural component of the time series, from the "noise," or "stochastic" component. We need to define some tools to help us separate the signal from the noise. Like the case of spatially related observations, useful tools will help us to investigate the variation among observations as a function of their separation distance, or "lag." Unlike spatially related observations, the temporal spacing of observations has only one dimension, time.

Basis for Tool Development

It seems reasonable that the statistical tools used for investigating a temporal series of observations ordered in time, $(z_1, z_2, z_3, ... z_N)$, should be based upon estimation of the variance of these observations as a function of their spacing in time. Such a tool is provided by the sample "autocovariance" function:

$$C_k = \frac{1}{N} \sum_{t=1}^{N-K} (z_t - \bar{z}) (z_{t+k} - \bar{z}) \, , \, k = 0, 1, 2, ... , K \qquad [8.1]$$

Here, $\bar{z}$ represents the mean of the series of N observations.

If we imagine that the time series represents a series of observations along a single dimension axis in space, then the astute reader will see a link between the covariance described by [8.1] and the variogram described by Equation [7.1]. This link is as follows:

$$\gamma(k) = C_0 - C_k \qquad [8.2]$$

The distance, k, represents the kth unit of time spacing, or lag, between time series observations.

A statistical series that evolves in time according to the laws of probability is referred to as a "stochastic" series or "process." If the true mean and autocovariance are unaffected by the time origin, then the stochastic process is considered to be "stationary." A stationary stochastic process arising from a *Normal*, or Gaussian, process is completely described by its mean and covariance function. The characteristic behavior of a series arising from *Normal* measurement "error" is a constant mean, usually assumed to be zero, and a constant variance with a covariance of zero among successive observations for greater than zero lag, $(k > 0)$. Deviations from this characteristic pattern suggest that the series of observations may arise from a process with a structural as well as a stochastic component.

Because it is the "pattern" of the autocovariance structure, not the magnitude, that is important, it is convenient to consider a simple dimensionless transformation of the autocovariance function, the autocorrelation function. The value of the autocorrelation, r_k, is simply found by dividing the autocovariance [8.1] by the variance, C_0:

$$r_k = \frac{C_k}{C_0}, \, k = 0, 1, 2 \, ... , K \qquad [8.3]$$

The sample autocorrelation function of the logarithm of PM_{10} concentrations presented in Figure 8.2 is shown below for the first 72 hourly lags. This figure illustrates a pattern that is much different from that characteristic of measurement error. It certainly indicates that observations separated by only one hour are highly related (correlated) to one another. The correlation, describing the strength of similarity in time among the observations, decreases as the distance separation, the lag, increases.

A number of estimates have been proposed for the autocorrelation function. The properties are summarized in Jenkins and Watts (2000). It is concluded that the most satisfactory estimate of the true kth lag autocorrelation is provided by [8.3].

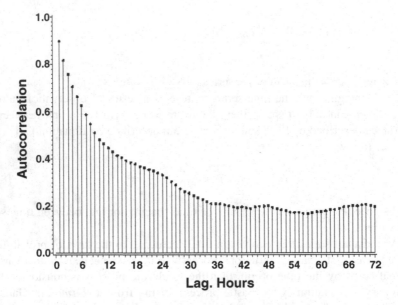

Figure 8.3   Autocorrelation of Ln Hourly PM$_{10}$ Observations, Liberty Borough Monitor, January–August 1993

It is necessary to discuss some of the more theoretical concepts regarding "general linear stochastic models" to assist the reader in appreciation of the techniques we have chosen for investigating and describing time series data. Few, if any, of the time series found in environmental studies result from stationary processes that remain in equilibrium with a constant mean. Therefore, a wider class of nonstationary processes called autoregressive-integrated moving average processes (ARIMA processes) must be considered. This discussion is not intended to be complete, but only to provide a background for the reader.

Those interested in pursuing the subject are encouraged to consult the classic work by Box et al. (1994), *Time Series Analysis Forecasting and Control*. Somewhat more accessible accounts of time series methodology can be found in Chatfield (1989) and Diggle (1990). An effort has been made to structure the following discussion of theory, nomenclature, and notation to follow that used by Box and Jenkins.

It should be mentioned at this point that the analysis and description of time series data using ARIMA process models is not the only technique for analyzing such data. Another approach is to assume that the time series is made up of sine and cosine waves with different frequencies. To facilitate this "spectral" analysis, a Fourier cosine transform is performed on the estimate of the autocovariance function. The result is referred to as the sample spectrum. The interested reader should consult the excellent book, *Spectral Analysis and Its Applications*, by Jenkins and Watts (2000).

Parenthetically, this author has occasionally found that spectral analysis is a valuable adjunct to the analysis of environmental times series using linear ARIMA models. However, spectral models have proven to be not nearly as parsimonious as parametric models in explaining observed variation. This may be due in part to the fact that sampling of the underlying process has not taken place at precisely the correct frequency in forming the realization of the time series. The ARIMA models appear to be less sensitive to the "digitization" problem.

ARIMA Models — An Introduction

ARIMA models describe an observation made at time t, say z_t, as a weighted average of previous observations, $z_{t-1}, z_{t-2}, z_{t-3}, z_{t-4}, z_{t-5}, \ldots$, plus the weighted average of independent, random "shocks," $a_t, a_{t-1}, a_{t-2}, a_{t-3}, a_{t-4}, a_{t-5}, \ldots$. This leads to the expression of the current observation, z_t, as the following linear model:

$$z_t = \phi_0 + \phi_1 z_{t-1} + \phi_2 z_{t-2} + \phi_3 z_{t-3} + \ldots + a_t - \theta_1 a_{t-1} - \theta_1 a_{t-2} - \theta_1 a_{t-3} - \ldots$$

The problem is to decide how many weighting coefficients, the ϕ's and θ's, should be included in the model to adequately describe z_t and secondly, what are the best estimates of the retained ϕ's and θ's. To efficiently discuss the solution to this problem, we need to define some notation.

A simple operator, the *backward shift operator* B, is extensively used in the specification of ARIMA models. This operator is defined by $Bz_t = z_{t-1}$; hence, $B^m z_t = z_{t-m}$. The inverse operation is performed by the *forward shift operator* $F = B^{-1}$ given by $Fz_t = z_{t+1}$; hence, $F^m z_t = z_{t+m}$. The *backward difference operator*, ∇, is another important operator that can be written in terms of B, since

$$\nabla z_t = z_t - z_{t-1} = (1 - B) z_t$$

The inverse of ∇ is the infinite sum of the binomial series in powers of B:

$$\nabla^{-1} z_t = \sum_{j=0}^{\infty} z_{t-j} = z_t + z_{t-1} + z_{t-2} + \ldots$$

$$= (1 + B + B^2 + \ldots) z_t$$

$$= (1 - B)^{-1} z_t$$

Yule (1927) put forth the idea that a time series in which successive values are highly dependent can be usefully regarded as generated from a series of *independent* "shocks," a_t. The case of the damped harmonic oscillator activated by a force at a random time provides an example from elementary physical mechanics. Usually, the shocks are thought to be random drawings from a fixed distribution assumed to be *normal* with zero mean and constant variance σ_a^2. Such a sequence of random variables a_t, a_{t-1}, a_{t-2}, ... is called *white noise* by process engineers.

A white noise process can be transformed to a nonwhite noise process via a *linear filter*. The linear filtering operation simply is a weighted sum of the previous realizations of the white noise a_t, so that

$$z_t = \mu + a_t + \Psi_1 a_{t-1} + \Psi_2 a_{t-2} + \cdots ,$$

$$z_t = \mu + \Psi(B)\, a_t$$

[8.4]

The parameter μ describes the "level" of the process, and

$$\Psi(B) = 1 + \Psi_1 B + \Psi_2 B^2 + \cdots$$

[8.5]

is the linear operator that transforms a_t into z_t. This linear operator is called the *transfer function* of the filter. This relationship is shown schematically.

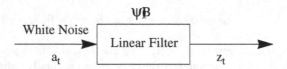

The sequence of weights ψ_1, ψ_2, ψ_3, ... may, theoretically, be finite or infinite. If this sequence is finite, or infinite and convergent, then the filter is said to be stable and the process z_t to be stationary. The mean about which the process varies is given by μ. The process z_t is otherwise nonstationary and μ serves only as a reference point for the level of the process at an instant in time.

Autoregressive Models

It is often useful to describe the current value of the process as a finite weighted sum of previous values of the process and a shock, a_t. The values of a process z_t, z_{t-1}, z_{t-2}, ... , taken at equally spaced times t, t – 1, t – 2, ... , may be expressed as deviations from the series mean forming the series $\tilde{z}_t$, $\tilde{z}_{t-1}$, $\tilde{z}_{t-2}$ ...; where $\tilde{z}_t = z_t - \mu$. Then

$$\tilde{z}_1 = \phi_1 \tilde{z}_{t-1} + \phi_2 \tilde{z}_{t-2} + \phi_p \tilde{z}_{t-p} + a_t$$

[8.6]

is called an autoregressive (AR) process of order p. An autoregressive operator of order p may be defined as

$$\phi(B) = 1 - \phi_1 B - \phi_2 B^2 - \dots - \phi_p B^p$$

Then the autoregressive model [8.6] may be economically written as

$$\phi(B)\,\tilde{z}_t = a_t$$

This expression is equivalent to

$$\tilde{z}_t = \Psi(B)\,a_t$$

with

$$\Psi(B) = \phi^{-1}(B)$$

Autoregressive processes can be either stationary or nonstationary. If the ϕ's are chosen so that the weights ψ_1, ψ_2, ... in $\Psi(B) = \phi^{-1}(B)$ form a convergent series, then the process is stationary.

Initially one may not know how many coefficients to use to describe the autoregressive process. That is, the order p in [8.6] is difficult to determine from the autocorrelation function. The pure autoregressive process has an autocorrelation function that is infinite in extent. However, it can be described in p nonzero functions of the autocorrelations.

Let ϕ_{kj} be the jth coefficient in an autoregressive process of order k, so that ϕ_{kk} is the last coefficient. The autocorrelation function for a process of order k satisfies the following difference equation where ρ_j represents the true autocorrelation coefficient at lag j:

$$\rho_j = \phi_{k1}\rho_{j-1} + \dots + \phi_{k(k-1)}\rho_{j-k+1} + \phi_{kk}\rho_{j-k} \qquad j = 1, 2 \ \dots, K \ [8.7]$$

This basic difference equation leads to sets of k difference equations for processes of order k (k = 1, 2, ... , p). Each set of difference equations are known as the *Yule-Walker* equations (Yule, 1927; Walker, 1931) for a process of order k. Note that the covariance of $(\tilde{z}_{j-k}a)$ vanishes when j is greater than k. Therefore, for an AR process of order p, values of ϕ_{kk} will be zero for k greater than p.

Estimates of ϕ_{kk} may be obtained from the data by using the estimated autocorrelation, r_j, in place of the ρ_j in the Yule-Walker equations. Solving successive sets of Yule-Walker equations (k = 1,2, ...) until ϕ_{kk} becomes zero for k greater than p provides a means of identifying the order of an autoregressive process. The series of estimated coefficients, ϕ_{11}, ϕ_{22}, ϕ_{33}, ... , define the *partial autocorrelation function*. The values of the partial autocorrelations ϕ_{kk} provide initial estimates of the weights ϕ_k for the autoregressive model Equation [8.6]

The clues used to identify an autoregressive process of order p are an autocorrelation function that appears to be infinite and a partial autocorrelation function which is truncated at lag p corresponding to the order of the process. To help us in deciding when the partial autocorrelation function truncates we can compare our estimates with their standard errors. Quenouille (1949) has shown that on the hypothesis that the process is autoregressive of order p, the estimated partial autocorrelations of order $p + 1$, and higher, are approximately independently distributed with variance:

$$\text{var}[\hat{\phi}_{kk}] \approx \frac{1}{N}$$

$$\text{var}[\hat{\phi}_{kk}] \approx \frac{1}{N} \quad k \geq p + 1$$

Thus the standard error (S.E.) of the estimated partial autocorrelation $\hat{\phi}_{kk}$ is

$$\text{S.E.}[\hat{\phi}_{kk}] \approx \frac{1}{\sqrt{n}} \quad k \geq p + 1$$

Moving Average Models

The autoregressive model [8.6] expresses the deviation $\tilde{z}_t = z_t - \mu$ of the process as a finite weighted sum of the previous deviations $\tilde{z}_t, \tilde{z}_{t-1}, \tilde{z}_{t-2} \ldots \tilde{z}_{t-p}$ of the process, plus a random shock, a_t. Equivalently as shown above the AR model expresses $\tilde{z}_t$ as an infinite weighted sum of the a's.

The finite moving average process offers another kind of model of importance. Here the $\tilde{z}_t$ are linearly dependent on a finite number q of previous a's. The following equation defines a moving average (MA) process of order q:

$$\tilde{z}_t = a_t - \theta_1 a_{t-1} - \theta_2 a_{t-2} - \ldots - \theta_q a_{t-q} \qquad [8.8]$$

It should be noted that the weights $1, \theta_1, \theta_2, \ldots, \theta_q$ need not have total unity nor need they be positive.

Similarly to the AR operator, we may define a moving average operator of order q by

$$\theta(B) = 1 - \theta_1 B - \theta_2 B^2 - \ldots \theta_q B^q$$

Then the moving average model may be economically written as

$$\tilde{z}_t = \theta(B) a_t.$$

This model contains $q + 2$ unknown parameters $\mu, \theta_1, \ldots, \theta_q, \sigma_a^2$, which have to be estimated from the data.

Identification of an MA process is similar to that for an AR process relying on recognition of the characteristic behavior of the autocorrelation and partial autocorrelation functions. The finite MA process of order q has an autocorrelation function which is zero beyond lag q. However, the partial autocorrelation function is infinite in extent and consists of a mixture of damped exponentials and/or damped sine waves. This is complementary to the characteristic behavior for an AR process.

Mixed ARMA Models

Greater flexibility in building models to fit actual time series can be obtained by including both AR and MA terms in the model. This leads to the mixed ARMA model:

$$\tilde{z}_t = \phi_1 \tilde{z}_{t-1} + \ldots + \phi_p \tilde{z}_{t-p} + a_t - \theta_1 a_{t-1} - \ldots - \theta_q a_{t-q} \qquad [8.9]$$

or

$$\phi(B) \tilde{z}_t = \theta(B) a_t$$

which employs $p + q + 2$ unknown parameters $\mu; \phi_1, \ldots, \phi_p; \theta_1, \ldots, \theta_q; \sigma_a^2$, that are estimated from the data.

While this seems like a very large task indeed, in practice the representation of actually occurring stationary time series can be satisfactorily obtained with AR, MA or mixed models in which p and q are not greater than 2.

Nonstationary Models

Many series encountered in practice exhibit nonstationary behavior and do not appear to vary about a fixed mean. The example of hourly PM_{10} concentrations shown in Figure 8.2 appears to be one of these. However, frequently these series do exhibit a kind of homogeneous behavior. Although the general level of the series may be different at different times, when these differences are taken into account the behavior of the series about the changing level may be quite similar over time. Such behavior may be represented by a generalized autoregressive operator $\varphi(B)$ for which one or more of the roots of the equation $\varphi(B) = 0$ is unity. This operator can be written as

$$\varphi(B) = \phi(B)(1 - B)^d$$

where $\phi(B)$ is a stationary operator. A general model representing homogeneous nonstationary behavior is of the form,

$$\varphi(B) z_t = \phi(B)(1 - B)^d z_t = \theta(B) a_t$$

or alternatively,

$$\phi(B) \ w_t = \theta(B) \ a_t \qquad [8.10]$$

where

$$w_t = \nabla^d z_t \qquad [8.11]$$

Homogeneous nonstationary behavior can therefore be represented by a model that calls for the dth difference of the process to be stationary. Usually in practice d is 0, 1, or at most 2.

The process defined by [8.10] and [8.11] provides a powerful model for describing stationary and nonstationary time series called an *autoregressive integrated moving average* (ARIMA) *process*, or order (p,d,q).

Model Identification, Estimation, and Checking

The first step in fitting an ARIMA model to time series data is the identification of an appropriate model. This is not a trivial task. It depends largely on the ability and intuition of the model builder to recognize characteristic patterns in the auto- and partial correlation functions. As always, this ability and intuition are sharpened by the model builder's knowledge of the physical processes generating the observations.

By way of illustration, consider the first three months of hourly PM_{10} concentrations from the Liberty Borough Monitor. This series is illustrated in Figure 8.4.

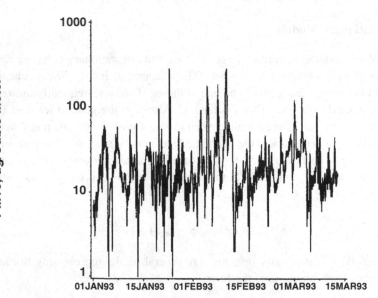

Figure 8.4   **Hourly PM_{10} Observations versus Time,**
Liberty Borough Monitor, January–March, 1993

Note that a logarithmic scale has been used on the PM_{10} concentration axis and a natural logarithmic transformation is applied to the data prior to initiating the analysis.

Figure 8.5 presents the autocorrelation function for the log-transform series, z_t. Note that the major behavior of this function is that of an exponential decay. However, there is the suggestion of the influence of a damped sine wave. Certainly, this behavior suggests a strong autoregressive component. This suggestion is also apparent in the partial autocorrelation function presented in Figure 8.6. The first partial autocorrelation coefficient is by far the most dominant feature. However, there is also a suggestion of the influence of a damped sine wave on this function after the first lag. Thus we have the possibility of a mixed autoregressive-moving average model. The dashed reference lines in each figure represent twice the standard error of the respective estimate.

There is no appropriate way to construct an ARIMA model. These models are usually constructed by "trial and error," conditioned with the experience and intuition of the analyst. Because of the strong suggestion of an autoregressive model in the example, an AR model of order 1 was used as a first try. This model is economically described by,

$$(1 - \phi_1 B) (z_t - \mu) = a_t \qquad [8.12]$$

Nonlinear estimates of the model parameters are obtain by the methods described by Box et al. (1994) (see also SAS, 1993). The derived estimates are

$$\mu = 2.835,$$

and

$$\phi_1 = 0.869.$$

These estimates may be evaluated by approximate t-tests (Box et al., 1994; SAS, 1993). However, the validity of these tests depend upon the adequacy of the model and the length of the series. Therefore, they should be used only with caution and serve more as a guide to the analyst than any determination of statistical significance.

Usually, the adequacy of the model is determined by looking at the residuals. Box et al. (1994) describe several procedures for employing the residuals in tests of deviations from randomness or "white noise." A chi-square test of the hypothesis that the model residuals are white noise uses the formula suggested by Ljung and Box (1978):

$$\chi_m^2 = n(n + 2) \sum_{k=1}^{m} \frac{r_k^2}{(n - k)}, \qquad [8.13]$$

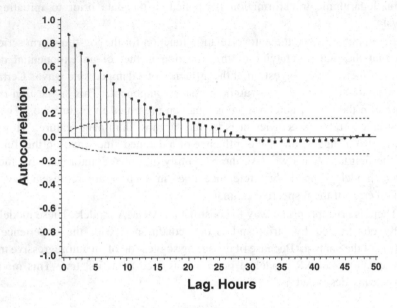

**Figure 8.5   Autocorrelation Function,
                 Log-Transformed Series**

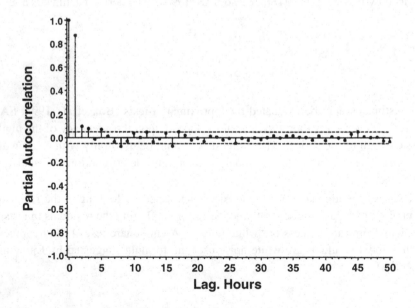

**Figure 8.6   Partial Autocorrelation Function,
                 Log-Transformed Series**

where

$$r_k = \frac{\sum_{t=1}^{n-k} a_t a_{t+k}}{\sum_{t=1}^{n} a_t^2}$$

and a_t is the series residual. Obviously, if the residual series is white noise, the r_k's are zero. The chi-square test applied to the residuals of our simple order AR 1 model indicates a significant departure of the model residuals from white noise.

In addition to assisting with a determination of model adequacy, the autocorrelations and partial autocorrelations of the residual series may be used to suggest model modifications if required. Figures 8.7 and 8.8 present the autocorrelation and partial autocorrelation functions of the series formed by the residuals from our estimated AR 1 model.

Note that both the autocorrelation and partial autocorrelation functions exhibit a behavior that in part looks like a damped sine wave. This suggests that a mixed ARMA model might be expected. However, there are precious few clues as to the number and order of model terms. There is the suggestion that something is affecting the system about every 15 hours and that there is a relationship among observations 3 and 6 hours apart. After some trial and error the following mixed ARMA model was found to adequately describe the data as indicated by the chi-square test for white noise:

$$(1 - \phi_1 B - \phi_{31} B^3 - \phi_6 B^6 - \phi_9 B^9)(z_t - \mu) = (1 - \theta_4 B^4 - \theta_{15} B^{15}) a_t, \qquad [8.14]$$

The estimated values for the model's coefficients are:

$$\mu = 2.828,$$

$$\phi_1 = 0.795,$$

$$\phi_3 = 0.103,$$

$$\phi_6 = 0.051,$$

$$\phi_9 = -0.066,$$

$$\theta_4 = 0.071, \text{ and}$$

$$\theta_{15} = -0.79.$$

This model provides a means of predicting, or forecasting, hourly values of PM_{10} concentration. Forecasts for the *median* hourly PM_{10} concentration and their 95 percent confidence limits are presented in Figure 8.9.

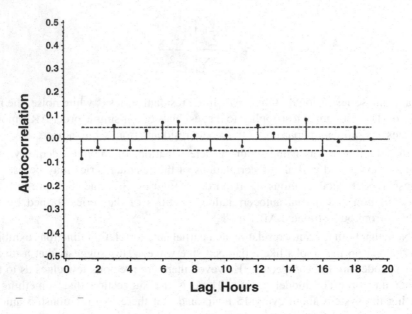

**Figure 8.7   Autocorrelation Function,
                Residual Series**

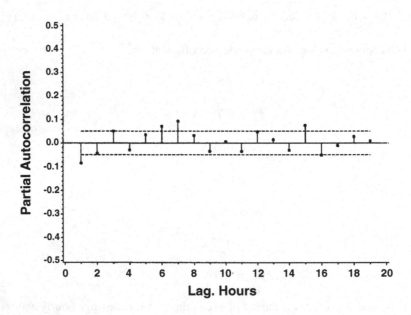

**Figure 8.8   Partial Autocorrelation Function,
                Residual Series**

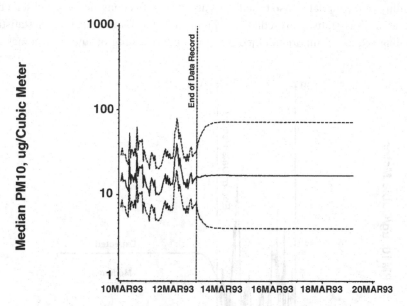

Figure 8.9   Forecasts of Hourly Medium PM$_{10}$ Concentrations

Note that there is little utility of forecasts made even a few hours beyond the end of the data record as the forecasts very rapidly become the predicted constant median value of the series.

The above model is a model for the natural logarithm of the hourly PM$_{10}$ concentration.   Simply exponentiating a forecast, $\hat{Z}$, of the series produces an estimate of the *median* of the series. This underpredicts the mean of the original series. If one wants to estimate the *expected* value, $\bar{Z}$, of the series the standard error of the forecast, s, needs to be taken into account.   On the assumption that the residuals from the model are normally distributed, the expected value is obtained from the forecast as follows:

$$\bar{Z} = e^{\left(\hat{Z} + \frac{S^2}{2}\right)}$$

[8.15]

The relationship between the median and expected value forecasts of the example series is shown in Figure 8.10.

It must be mentioned that there is more than one ARIMA model that may fit a given time series equally as well. The key is to find that model that best meets the needs of the user. The reader is reminded that "... all models are wrong but some are useful" (Box, 1979). The utility of any particular model depends largely upon how well it accomplishes the task for which it was designed.   If the desire is only to forecast future events, then the utility will become evident when these future observations come to light. However, as frequently is the case, the task of the modeling exercise is to identify factors influencing environmental observations. Then the utility of the model is also based in its ability to reflect engineering and scientific logic as well as statistical prediction. Frequently the forensic nature of statistical modeling is a more important objective than the forecasting of future outcomes.

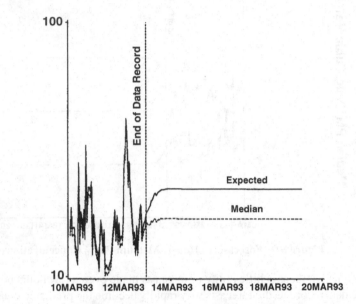

Figure 8.10   Expected and Median Forecasts of
PM$_{10}$ Concentrations

An example of this is provided by the PM$_{10}$ measurements made at the Allegheny County Liberty Borough monitoring site between March 10 and March 17, 1995. Figure 8.11 presents this hourly measurement data. The dashed line give the level of the hourly standard. During this time span several exceedances of the hourly 150 µg/m^3 air quality standard occurred. Also, during this period six nocturnal temperature inversions of a strength greater than 4 degrees centigrade were recorded and industrial production in the area was reduced in accordance with the Allegheny County air quality episode abatement plan.

It is interesting to look at a three-dimensional scatter diagram of PM$_{10}$ concentrations as a function of wind speed and direction for the Liberty Borough monitor site. This is given in Figure 8.12. Note that there is an obvious difference in PM$_{10}$ associated with both wind direction and speed. Traditionally, urban air quality monitoring sites are located so as to monitor the impact of one or more sources. The Liberty Borough monitor is no exception. A major industrial source is upwind of the monitor when the wind direction is from SSW to SW. The alleged impact of this source is evident with the higher PM$_{10}$ concentrations associated with winds from 200 to 250 degrees. This directional influence is obviously dampened by wind speed.

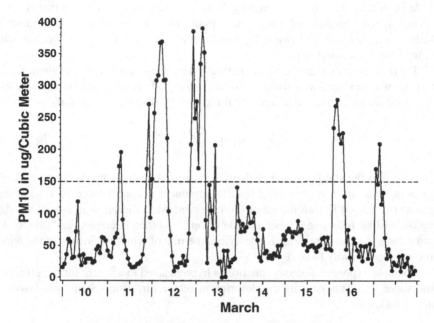

Figure 8.11   Hourly PM_{10} Observations,
Liberty Borough Monitor, March 10–17, 1995

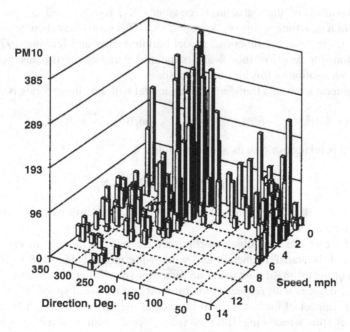

Figure 8.12   Hourly PM_{10} Observations versus Wind Direction and Speed,
Liberty Borough Monitor, March 10–17, 1995

In order to account for any "base load" associated with this major source, a wind direction-speed "windowing" filter was hypothesized and its parameters estimated. The hypothesized filter has two components, one to account for wind direction and one to account for wind speed.

The direction filter can be mathematically described very nicely by one of those functions whose utility was always in doubt during differential equations class, the hyperbolic secant (Sech). The functional form of the direction filter, $dirf_t$, is

$$dirf_t = Sech\frac{\pi K_1(\delta_t - \Delta_0)}{180} \qquad [8.16]$$

Here, δ_t is the wind direction in degrees measured from the north at time t. Sech ranges in value from approaching 0 as its argument becomes large to 1 when its argument is zero. Therefore, when the observed wind direction δ_t equals Δ_0 the window will be fully open, have a value of one. Δ_0 then becomes the "principal" wind direction. The parameter K_1 describes the rate of window closure as the wind direction moves away from Δ_0.

A simple exponential decay function is hypothesized to account for the effect of wind speed, u. This permits the description of the direction-speed "windowing" filter as follows:

$$x_t = K_3 e^{-K_2 u_t} Sech\frac{\pi K_1(\delta - \Delta_0)}{180} \qquad [8.17]$$

Given values of the "structural" constants K_1, K_2, K_3, and Δ_0 permits the formation of a new time series, $x_1, x_2, x_3, \dots, x_t$. This series may then be used as an "input" series in "transfer function" model building (Box and Jenkins, 1970). The resulting transfer function model and structural parameter estimates permit the forensic investigation of this air quality episode.

The general form of a transfer function model with one input series is given by

$$(1 - \delta_1 B - \delta_2 B^2 - \dots - \delta_r B^r)(Y_t - \mu) = (\omega_0 - \omega_1 B - \dots - \omega_r B^r) X_{t-b} + N_t \qquad [8.18]$$

Rewriting this relationship in its shortened form,

$$Y_t = \mu + \delta^{-1}(B)\omega(B) X_{t-b} + N_t \qquad [8.19]$$

where $N_t = \varphi^{-1}(B)\theta(B) a_t$ represent the series "noise" in terms of an ARMA model of the random shocks.

ARIMA and other nonlinear techniques are used iteratively to estimate the parameters of the transfer function and windowing models. Figure 8.13 illustrates the results of the estimation on the wind direction-speed filter. If the wind direction is from 217 degrees with respect to the monitor and the wind speed is low, the full "base load" impact of the source will be seen at the Liberty Borough Monitor. In other words, the windowing filter is fully "open" with a value of one. The windowing filter closes, has smaller and smaller values, as either the wind direction moves away from 217 degrees or the wind speed increases.

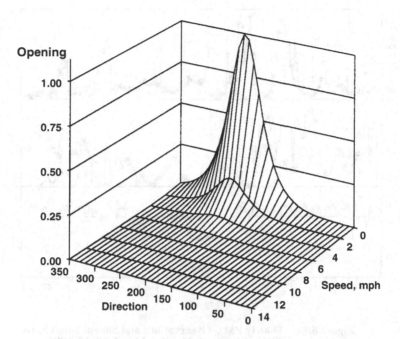

**Figure 8.13   Wind Direction and Speed Windowing Filter,
Liberty Borough Monitor, March 10–17, 1995**

Note that the scatter diagram in Figure 8.12 indicates that PM_{10} concentrations at Liberty Borough are also elevated when the wind direction is from the north and possibly east. These "northern" and "eastern" elevated concentrations appear to be associated with low wind speed.   This might suggest that wind direction measurement at the site is not accurate at low wind speed and is misleading. However, if the elevated concentrations in the "northern" and "eastern" directions were a result of an inability to measure wind direction at low wind speeds, a uniform pattern of PM_{10} concentration would be expected at low wind speeds. Obviously, this is not the case. This leads to the conclusion that other sources may exist north and east of the Liberty Borough monitor site. These sources could be quite small in terms of PM_{10} emissions, but they do appear to have a significant impact on PM_{10} concentrations measured at Liberty Borough.

Figure 8.14 illustrates some potentially interesting relationships between PM_{10} concentrations at the Liberty Borough monitor and other variables considered in this investigation. The top panel presents the hourly PM_{10} concentrations as well as the strength and duration of each nocturnal inversion. Note that PM_{10} generally increases during the inversion periods. The middle panel shows the magnitude of the directional windowing filter and wind speed.

Comparing the data presented in the top and middle panels, it is obvious that (1) the high PM_{10} values correspond to an "open" directional filter (value close to 1) and low wind speeds, and (2) this correspondence generally occurs during periods of

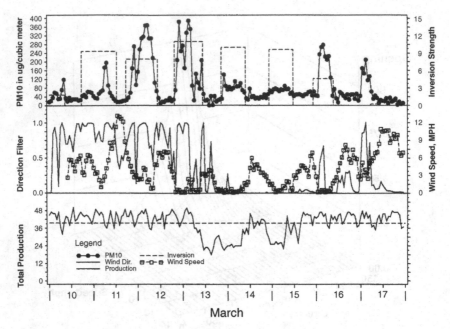

**Figure 8.14   Hourly PM$_{10}$ Observations and Salient Input Series,
Liberty Borough Monitor, March 10–17, 1995**

inversion. The notable exception is March 17. Even here the correspondence of higher PM$_{10}$ and wind direction and speed occurs during the early hours of March 17 when the atmospheric conditions are likely to be stable and not support mixing of the air.

The bottom panel presents the total production index as a surrogate for production at the principal source. The decrease in production on March 13 and subsequent return to normal level is readily apparent. It is obvious from comparison of the bottom and middle panels that the decrease in production corresponds with a closing of the direction window (low values). Thus, any inference regarding the effectiveness of decreasing production on reducing PM$_{10}$ levels is totally confounded with any effect of wind direction.

One should not expect that every "high" PM$_{10}$ concentration will have a one-to-one correspondence open directional window and low wind speed. This is because the factors influencing air quality measurements do not necessarily run on the same clock as that governing the making of the measurement. Because air quality measurements are generally autocorrelated, they remember where they have been. If an event initiates an increase in PM$_{10}$ concentration at a specific hour, the next hour is also likely to exhibit an elevated concentration. This is in part because the initiating event may span hours and in part because the air containing the results of the initiating event does not clear the monitor within an hour. The latter is particularly true during periods of strong temperature inversions.

Summarizing, a "puff" of fine particulate matter from the principal source will likely impact the monitoring site if a light wind is blowing from 217 degrees during a period of strong temperature inversion. In other words the wind direction-speed window is fully open and the "storm window" associated with temperature inversions is also fully open. If the storm window is partially closed, i.e., a weak temperature inversion, permitting moderate air mixing, the impact of the principal source will be moderated.

Letting S_t represent the strength of the temperature inversion in degrees at time t, the inversion "storm window" can be added to the wind direction-speed window as a simple linear multiplier. This is illustrated by the following modification of Equation 8.17:

$$x_t = K_3 \frac{S_t}{11} e^{-K_2 u_t} \text{Sech} \frac{\pi K_1 (\delta_t - \Delta_d)}{180} \qquad [8.20]$$

Building the transfer function model between PM_{10} concentration, Y_t, and the inversion wind direction-speed series, X_t, relies on identification of the model form the cross-correlation function between the two series. It is convenient to first "prewhiten" the input series by building an ARIMA model for that series. The same ARIMA model is then applied to the output series as a prewhitening transformation. Using the cross-correlation function (Figure 8.15) between the prewhitened input series and output series one can estimate the orders of the right- and left-hand side polynomials, r and s, and backward shift b in Equation 8.18.

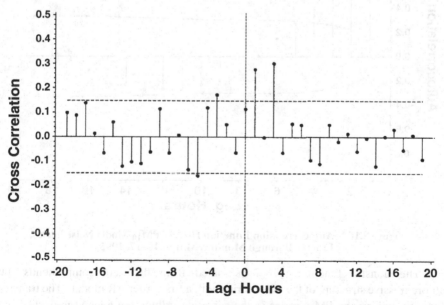

Figure 8.15   Cross Correlations Prewhitened Hourly PM$_{10}$ Observations and Input Series, Liberty Borough Monitor, March 10–17, 1995

Box et al. (1994) provide some general rules to help us. For a model of the form 6.18 the cross-correlations consist of

(i)   b zero values $c_0, c_1, \dots, c_{b-1}$;

(ii) a further $s - r + 1$ values $c_b, c_{b+1}, \dots, c_{b+s-r}$, which follow no fixed pattern (no such values occur if $s < c$);

(iii) values c_j with $j \geq b + s - r + 1$ which follow the pattern dictated by an rth order difference equation that has r starting values $c_{b+s}, \dots, c_{b+s-r+1}$. Starting values c_j for $j < b$ will be zero. These starting values are directly related to the coefficients $\delta_1, \dots, \delta_r$ in Equation 8.18.

The "noise" model must also be specified to complete the model building. This is accomplished by identifying the noise model from the autocorrelation function for the noise as with any other univariate series. The autocorrelation function for the noise component is given in Figure 8.16.

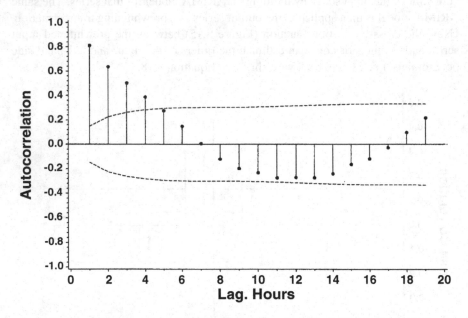

Figure 8.16   Autocorrelation Function Hourly PM$_{10}$ Model Noise Series, Liberty Borough Monitor, March 10–17, 1995

The transfer function model estimated from the data comprehends the autoregressive structure of the noise series with a first-order AR model. The transfer function linking the PM$_{10}$ series to the series describing the alleged impact of the principal source filtered by meteorological factor window has a one-hour back shift (b = 1) and numerator term of order three (s = 3). Using the form of Equation 8.18, this model is described as follows:

$$Y_t = 65.14 + (18.333 - 4.03B + 187.64B^2 - 27.17B^3) X_{t-1} + \frac{1}{(1 - 0.76B)} a_t \qquad [8.21]$$

This model accounts for 76 percent of the total variation in PM_{10} concentrations over the period. Much of the unexplained variation appeared to be due to a few large differences between the observed and predicted PM_{10} values. It can be hypothesized that a few isolated, perhaps fugitive emission, events may be providing a "driving" force for the observed unexplained variation appearing as large differences between the observed and predicted PM_{10} concentration behavior. The occurrence of such events might well correspond to the large positive differences between the observed and predicted PM_{10} concentrations.

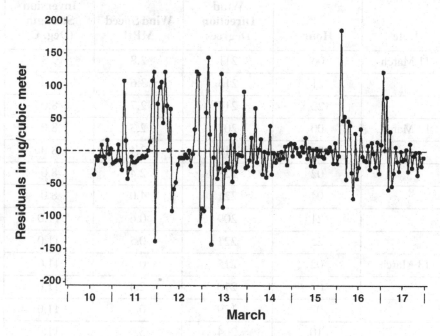

**Figure 8.17   Hourly PM_{10} Model [6.21] Residuals,
Liberty Borough Monitor, March 10–17, 1995**

A new transfer function model was constructed for the March 1995 episode including the 19 hypothesized "events" listed in Table 8.1. These events form a binary series, I_t, which has the value of one when the event is hypothesized to have occurred and zero otherwise. Figure 8.18 presents the model's residuals. This model given by Equation 8.22 accounted for 90 percent of the total observed variation in PM_{10} concentration at the Liberty Borough monitor:

$$Y_t = 29.47 + \frac{(1.31 - 0.04B - 0.22B^2)}{(1 - 0.78B)} I_t +$$

$$(44.31 - 15.11B + 199.34B^2 - 106.89B^3) X_{t-1} + \frac{1}{(1 - 0.79B)} a_t \qquad [8.22]$$

The binary variable series, I_t, is an "intervention" variable. Interestingly, Box and Tiao (1975) were the first to propose the use of "intervention analysis" for the investigation of environmental studies. Their environmental application was the analysis of the impact of automobile emission regulations on downtown Los Angeles ozone concentrations.

Table 8.1
Hypothesized Events

Date	Hour	Wind Direction Degrees	Wind Speed MPH	Inversion Strength (Deg. C)
11 March	06	211	2.8	9.3
	21	213	3.6	8.0
	22	217	2.7	8.0
12 March	00	207	2.3	8.0
	01	210	2.4	8.0
	02	223	2.4	8.0
	04	221	4.0	8.0
	21	209	0.6	11.0
	22	221	0.5	11.0
13 March	02	215	0.7	11.0
	03	210	2.4	11.0
	07	179	0.5	11.0
	10	204	3.3	11.0
	22	41	0.1	10.0
14 March	04	70	0.2	10.0
16 March	02	259	0.2	4.6
	04	245	0.5	4.6
17 March	01	201	5.2	0.0
	03	219	4.0	0.0

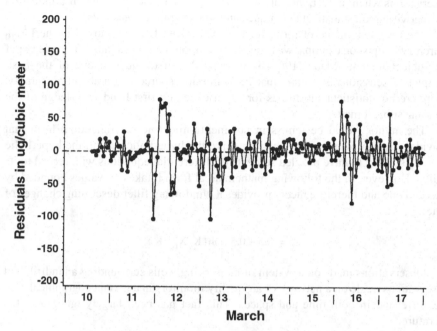

Figure 8.18 **Hourly PM_{10} Model [6.22] Residuals,**
Liberty Borough Monitor, March 10–17, 1995

The "large" negative residuals are a result of the statistical model not being able to adequately represent very rapid changes in PM_{10}. Negative residuals result when the predicted PM_{10} concentration is greater than the observed. They may represent situations where the initiating event was of sufficiently minor impact that its effect did not extend for more than the hourly observational period or a sudden drastic change occurred in the input parameters. The very sudden change in wind direction at hour 23 on March 11 is an example of the latter.

The deviations between observed and predicted PM_{10} concentrations for the transfer function employing the 19 hypothesized "events" are close to the magnitude of "measurement" variation. These events are a statistical convenience to improve the fit of an empirical model. There is, however, some allegorical support for their correspondence to a fugitive emission event.

Epilogue

The examples presented in this chapter have been limited to those regarding air quality. Other examples of environmental time series are found in waste water discharge data, groundwater quality data, stack effluent data, and analyte measurements at a single point in a water body to mention just a few. These examples were mentioned at the beginning of this chapter, but the point bears repeating. All too often environmental data are treated as statistically independent observations when, in fact, they are not. This frequently leads to the misapplication of otherwise good statistical techniques and inappropriate conclusions.

The use, and choice of form, for the windowing filter employed in the PM_{10} March 1995 episode example was conditioned upon the investigator's knowledge of the subject matter. Slade (1968) is perhaps the first, and still one of the best, complete discussions of factors that link emission source and ambient air quality. There are no statistical substitutes for experience and first-hand knowledge of the germane subject matter.

The author would be remiss in not mentioning another functional form that provides a useful windowing function. Like the hyperbolic secant, the hyperbolic tangent is likely to have puzzled students of differential equations with regard to its utility. However, in the following mathematical form it takes on values bounded by zero and one and therefore nicely provides a windowing filter describing changes of state:

$$Z_t = 0.5 + 0.5 \tanh(K_1 X_t - K_0)$$

Observations made on a system of monitoring wells surrounding a landfill, and observations made on a network of air quality monitors in a given geographical area are correlated in both time and space. This fact has been largely ignored in the literature.

References

Box, G. E. P., 1979, "Robustness in the Strategy of Scientific Model Building," *Robustness in Statistics*, eds. R. L. Launer and G. N. Wilkinson, Academic Press, pp. 201–236.

Box, G. E. P., Jenkins, G. M., and Reinsel, G. C., 1994, *Time Series Analysis, Forecasting and Control*, 3rd ed., Prentice Hall, Englewood Cliffs, NJ.

Box, G. E. P. and Tiao, G. C., 1975, "Intervention Analysis with Applications to Economic and Environmental Problems," *Journal of the American Statistical Association*, 70(349): 70–79.

Chatfield, C., 1989, *The Analysis of Time Series: An Introduction*, Chapman and Hall, New York.

Cleveland, W. S., 1972, "The Inverse Autocorrelations of a Time Series and Their Applications," *Technometrics*, 14: 277.

Deming, W. E., 1950, *Some Sampling Theory*, John Wiley and Son, New York, pp. 502–203.

Diggle, P. J., 1990, *Time Series: A Biostatistical Introduction*, Oxford University Press, New York.

Jenkins, G. M. and Watts, D. G., 2000, *Spectral Analysis and Its Applications*, Emerson Adams Pr, Inc.

Ljung, G. M. and Box, G. E. P., 1978, "On a Measure of Lack of Fit in Time Series Models," *Biometrika*, 65: 297–303.

Quenouille, M. H., 1949, "Approximate Tests of Correlation in Time Series," *Journal of the Royal Society*, B11.

SAS Institute Inc., 1993, *SAS/ETS User's Guide, Version 6, Second Edition*, Cary, NC, pp. 99–182.

Slade, D. H. (ed.), 1968, *Meteorology and Atomic Energy*, Technical Information Center, U.S. Department of Energy.

Walker, G., 1931, "On Periodicity in Series of Related Terms," *Proceedings of the Royal Society*, A131.

Yule, G. U., 1927, "On a Method of Investigating Periodicities in Disturbed Series, with Special Reference to Wölfer's Sunspot Numbers," *Philosophical Transactions*, A226.

Index

Printed in the United States
by Baker & Taylor Publisher Services